MEPTEC Semiconductor Roadmaps Symposium 2013

A Collaborative Update from Standards Bodies, Industry Groups, and the Entire Supply Chain

MEPTEC Symposium Proceedings Number 56

Santa Clara, California, USA
24 September 2013

ISBN: 978-1-62993-805-9

Printed from e-media with permission by:

Curran Associates, Inc.
57 Morehouse Lane
Red Hook, NY 12571

Some format issues inherent in the e-media version may also appear in this print version.

Copyright© (2013) by MEPTEC
All rights reserved.

Printed by Curran Associates, Inc. (2014)

For permission requests, please contact MEPTEC
at the address below.

MEPTEC
P.O. Box 222
Medicine Park, Oklahoma 73557

Phone: 650-714-1570
Fax: 1-866-424-0130

info@meptec.org

Additional copies of this publication are available from:

Curran Associates, Inc.
57 Morehouse Lane
Red Hook, NY 12571 USA
Phone: 845-758-0400
Fax: 845-758-2634
Email: curran@proceedings.com
Web: www.proceedings.com

MEPTEC Semiconductor Roadmaps Symposium 2013

A Collaborative Update from
Standards Bodies, Industry Groups,
and the Entire Supply Chain

MEPTEC Symposium Proceedings Number 56

Santa Clara, California, USA
24 September 2013

TABLE OF CONTENTS

PANEL ONE: PRODUCT DRIVERS
Moderator: Joel Camarda

Product Drivers: An OEM Perspective 1
Mudasir Ahmad

Package Drivers: What Does LSI Do? 4
Farshad Ghahghahi

CMOS Imager Sensor (CIS) Packaging & Product Drivers 7
Larry Kinsman

Mobile Deman Drives Growth in Advanced Packaging 9
Tom Strothmann

Drivers & Packaging Challenges at SanDisk 11
Suresh Upadhyayula

Panel One Summary 13
Paul Werbaneth, Mudasir Ahmad, Farshad Ghahghahi, Larry Kinsman, Tom Strothmann, Suresh Upadhyayula

PANEL TWO: MANUFACTURING DRIVERS IN SEMICONDUCTOR ROADMAPS
Moderator: Jeffrey C. Demmin

Panel Two Introduction 19
Jeffrey C. Demmin, Richard Crisp, Javier DeLaCruz, Ron Huemoller, Raj Master, Dongkai Shangguan

Invensas Corporation 21
Richard Crisp

Manufacturing Drivers in Semiconductor Roadmaps 27
Javier DeLaCruz

SVP, Adv. Product / Platform Develop 30
R. Huemoeller

Microsoft Hardware Teams: 30 Years of Innovation 33
Raj Master

3D Microelectronics Packaging: Technology Development & Commercialization 36
Dongkai Shangguan

Panel Two Summary 39
Linda Matthew, Richard Crisp, Javier DeLaCruz, Ron Huemoller, Raj Master, Dongkai Shangguan

KEYNOTE

The Collaboration Engine: Enabling Innovation in Microelectronics 44
Karen Savala

PANEL THREE: ELECTRICAL PERFORMANCE REQUIREMENTS TO MEET EMERGING INTERCONNECT STANDARDS
Moderators: Ivor Barber, John Xie

Panel Three Introduction 56
Ivor Barber, Tom Gregorich, Brad Griffin, Anthony Torza, Abe Yee

Electrical Requirements to Meet Emerging Interconnect Standards 58
Tom Gregorich

IC Package Design Trends Electrical Assessment – Early and Often 62
Brad Griffin

Next Generation FPGA Packaging Trends 65
Anthony Torza

GPU Roadmap 68
Abe Yee

Panel Three Summary 70
Ivor Barber, Tom Gregorich, Brad Griffin, Anthony Torza, Abe Yee

PANEL FOUR: THE IMPORTANCE OF INDUSTRY ORGANIZATION COLLABORATION
Moderator: Phil Marcoux

Panel Four Introduction..75
Phil Marcoux, Dieter Bergman, Herb Reiter, Kenneth Potts, Jasbir Bath, Paul Trio

Coalition for Advancement of MicroElectronic Systems Technology - "CAMEST" Organization and Plans..76
Dieter Bergman, Denny Fritz

The Importance of Industry Organization Collaboration..81
Herb Reiter

3D IC Working Group Objective...84
Kenneth Potts

The Importance of Industry Organization Collaboration..90
Jasbir Bath

SEMI Standards Program Overview..92
Paul Trio

Panel Four Summary..94
Phil Marcoux, Dieter Bergman, Herb Reiter, Kenneth Potts, Jasbir Bath, Paul Trio

Author Index

About MEPTEC

MEPTEC is a trade association of semiconductor companies and professionals involved in the manufacturing, packaging, assembling and testing of integrated circuits. Since its inception over 30 years ago, MEPTEC has provided a forum for the semiconductor industry to learn and exchange ideas through our monthly luncheons, conferences, and our quarterly publication, the MEPTEC Report. With the support of an Advisory Board consisting of individuals from all segments of the industry, MEPTEC has, over the years, kept current not just with semiconductor industry developments, but has expanded its scope to cover relevant industry segments such as MEMS and medical electronics. For more information about MEPTEC events and membership, please visit www.meptec.org.

P.O. Box 222, Medicine Park, OK 73557
Tel: 650-714-1570   Fax: 1-866-424-0130
Email: info@meptec.org

Visit the MEPTEC website at www.meptec.org for more information.

 **2013 SEMICONDUCTOR ROADMAPS SYMPOSIUM**

A SPECIAL THANKS TO OUR MEDIA SPONSORS

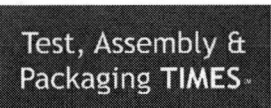

Tuesday, September 24, 2013 • Biltmore Hotel • Santa Clara, California

 **MEPTEC** 2013 **SEMICONDUCTOR ROADMAPS SYMPOSIUM**

A SPECIAL THANKS TO OUR GOLD SPONSOR

FAB OWNERS ASSOCIATION

Tuesday, September 24, 2013 • **Biltmore Hotel** • **Santa Clara, California**

"Collaborate For Success"

Collaboration is the key to efficiency & profitability in today's challenging semiconductor industry. The unique goal of the FOA is to enable our nearly 100 members - semiconductor & MEMS manufacturers along with their suppliers - to share their strengths & resources for the benefit of all.

Learn more at www.waferfabs.org

We would like to thank our sponsors and exhibitors for supporting this event!

PO Box 222, Medicine Park, OK 73557
Tel: 650-714-1570   Fax: 866-424-0130
www.meptec.org

 **MEPTEC** 2013 SEMICONDUCTOR ROADMAPS SYMPOSIUM

BIOGRAPHIES

SESSION CHAIRS

Ivor Barber graduated from Napier University in Edinburgh, Scotland in 1981 with a Bachelors degree in Technology. He has worked in package assembly and design at National Semiconductor, Fairchild Semiconductor and VLSI Technology. Ivor spent 23 years at LSI Corporation in Milpitas in various Engineering and Management positions in Assembly, Package Characterization and Package Design. Ivor is currently Senior Director of Package Technology Development at Xilinx. Ivor holds 13 US patents related to package design.

Joel Camarda is an industry consultant, concentrating on manufacturing operations management and is also a Sr. Member Technical Staff for Amonix, a leading CPV system supplier. He well known in the international packaging community via his work history of 30+ years in the USA and Asia. He has been active in IMAPS and is an advisor for MEPTEC. Joel has held several executive management positions: VP Operations at Sipex/Exar, President of K&S Flip Chip Technology, and Director of Worldwide Assembly and Packaging at Cypress Semiconductor. He started his career at National Semiconductor.

Jeffrey C. Demmin is the Director of OEM Marketing at STATS ChipPAC, a leading provider of semiconductor assembly and test services. Before that he worked at Tessera Technologies in marketing, corporate development, and IP acquisition. He was previously the Editor-in-Chief of Advanced Packaging magazine and Senior Technical Editor of Solid State Technology magazine. His career started in semiconductor package design at National Semiconductor, and a sequence of engineering roles at nCHIP, Seagate, and Textron followed that. Jeff earned a bachelor's degree in Physics from Princeton University and a master's degree in Materials Science from Stanford University. He has been awarded five patents and a gold medal from the American Society of Business Publication Editors (ASPBE).

Phil Marcoux is one of many SMT and IC Packaging Pioneers. In 2007 he was named "The Father of US SMT" by the IPC. In 1981 he founded, AWI, the first US Company devoted exclusively to SMT which was later acquired by SCI Systems. In 1992 he founded, ChipScale, one of the first Wafer Level Packaging companies which developed a portfolio of over 36 patents. The patents are now the cornerstone of the camera modules commonly found in the current cell phones, computers, and games. Today, Phil is an active Business Development consultant in the area of 2.5D and 3D IC packaging infrastructure, design, and assembly.

John Yuanlin Xie, Ph.D. has been with Altera Corporation for 14 years. His current position is Director of Packaging Technology Research and Development. Prior to Altera, he was a technology development manager at Prolinx Labs. Dr. Xie graduated from Department of Physics, Peking University, and holds a Ph.D. Degree in Physics from Institute of Physics, Chinese Academy of Sciences and Post Doctoral from Department of Physics, University of California, Berkeley and Lawrence Berkeley Laboratory. Dr. Xie has 24 published patents with 10 more pending; and over 40 academic and technical publications. He is the President of Chinese Institute of Engineers USA, San Francisco Bay Area Chapter.

KEYNOTE SPEAKER

Karen Savala is president of SEMI Americas and is responsible for the association's Americas programs, including events, products and services. She is responsible for relationships with SEMI members as well as industry, government and academia in the region. Savala joined SEMI in 1984 and has served in numerous managerial and executive roles, including positions in International Standards, executive programs, and outreach and membership. She established the "Voice of the Customer" program which helped drive product and service improvements to improve SEMI member satisfaction. Savala earned a business management and communications degree from San Jose State University in San Jose, California.

(continued)

Tuesday, September 24, 2013 • **Biltmore Hotel** • **Santa Clara, California**

 2013 SEMICONDUCTOR ROADMAPS SYMPOSIUM

PARTICIPANTS / PANELISTS

Mudasir Ahmad is a Distinguished Engineer at Cisco Systems, Inc. He has been involved with microelectronics packaging design and reliability analysis for 15 years. He received his Bachelors from Ohio University and his M.S. degree in Mechanical Engineering from Georgia Institute of Technology. He is currently pursuing his Masters in Management Science & Engineering at Stanford. At Cisco, Mudasir is leading the Center of Excellence for Numerical Analysis, developing new analytical algorithms, experimental design and reliability characterization of next generation 3D packaging, System-in-Package Modules and Silicon Photonics. Mudasir is also involved with implementing new technologies like Cloud Computing and Big Data Analytics to streamline Supply Chain Operations at Cisco. Outside of Cisco, he is the Vice Chair of the Silicon Valley Chapter of the Components Packaging and Manufacturing Technology (CPMT) Society of the IEEE. Mudasir has over 25 publications on microelectronic packaging, two book chapters, and 7 US Patents.

Jasbir Bath is a Principal Engineer in the Assembly Technology Group at IPC. He was the Corporate Lead Engineer with Solectron Corporation and Flextronics International for 10 years with a role involving tin-lead and lead-free solder process development. Previously he was a Technical Officer at ITRI (International Tin Research Institute/ Tin Technology) Ltd in the U.K. He holds BS and MS degrees in Materials Science from the University of Manchester/ UMIST in England, U.K. Jasbir is the editor of the Springer Publications book: Lead-free Soldering and co-editor of 2 IEEE-Wiley books: Lead-free Electronics, iNEMI Projects Lead to Successful Manufacturing and Lead-free Solder Process Development. He has also co-authored the Lead-free Manufacturing chapter of the IEEE-Wiley book: Lead-free Electronics.

Dieter Bergman, IPC Ambassador of Technical Programs, has worked in the Electronics Industry for over 55 years. He began his career in 1956 as an electromechanical designer for Philco-Ford in Philadelphia, PA. He became supervisor of the printed circuit design group in 1967, and joined the company's advanced technology group where he specialized in printed circuit computer-aided design. He joined IPC as Technical Director in 1974, and was instrumental in the development of many of the IPC Design and Performance standards. At present he is does contract work for IPC on related topics of Data Transfer, Computer Formats and Design related Implementation projects.

Richard Crisp is Vice president and Chief Technologist at Invensas Corporation, a wholly owned subsidiary of Tessera Technologies, Inc. (Nasdaq: TSRA). Crisp is responsible for product strategy, development and promotion of Invensas' semiconductor packaging technologies with a particular focus on advanced system integration structures. Crisp has worked for Motorola, Intel, MIPS and Rambus. Crisp began his design career as a key circuit designer of the Motorola MC68000 Microprocessor. He led the design of the first two generations of RAMBUS DRAMs, introducing a number of fundamental inventions now in common usage throughout the DRAM industry. Crisp was the Program Committee Chairman and Vice Chair for IEEE's International Solid State Circuits Conference (ISSCC 2000 and ISSCC 1999) and was the Memory Subcommittee Chairman before. Crisp has authored numerous peer-reviewed papers for IEEE and SPIE journals and conferences. Crisp has been awarded 35 US Patents and has 43 published US Patent applications. He received his Bachelor's degree, with Cum Laude honors, in Electrical Engineering from Texas A&M University in 1976.

Javier DeLaCruz is the senior director of engineering at eSilicon Corporation. He is responsible for foundry, packaging, signal integrity, product and test engineering. He has also lead the 2.5D and 3D package/system initiatives branded as MoZAIC™ devices. DeLaCruz has over 17 years of experience in semiconductor packaging and is a senior member of IEEE. He has extensive knowledge in analog design, high-speed digital design, thermal and electrical simulation, and package assembly.

Farshad Ghahghahi graduated from Southwest Minnesota State University in Minnesota, in 1982 with a degree in Mechanical Engineering. He has worked at Tandem Computers in development of component and system packaging for 14 years. Farshad has spent the last 16 years at LSI Corporation in San Jose, CA in various Engineering and Management positions in Package Design. Farshad is currently Director of Package Design and Characterization at LSI Corporation. Farshad holds 17 US patents related to component package design, and has presented papers in the area of Package Co-Design and new Developments.

Tom Gregorich is Director of Package Signal Integrity and Thermal Design Optimization at Broadcom. Previously Mr. Gregorich held positions at MediaTek, where he developed one of the first commercial products to use coreless Cu-pillar technology, and Qualcomm, where he established the Package Engineering Department and for 12 years led the development of Qualcomm's

Tuesday, September 24, 2013 • Biltmore Hotel • Santa Clara, California

small form-factor package portfolio including 0.4mm CSP, molded-laser POP and package-in-package. A large portion of the Qualcomm package portfolio is 3D, and Mr. Gregorich established the TSV program at Qualcomm with IMEC. Prior to his position at Qualcomm, Mr. Gregorich worked for Motorola and had assignments in the Semiconductor Products Sector and Corporate Research, both in the United States as well as Japan and China. Mr. Gregorich has a BS in Mechanical Engineering from Bradley University, an MBA from Northern Illinois University and is a Senior Member of IEEE.

Brad Griffin is a product marketing director, High-Speed Solutions for SiP, IC Packaging, and PCB in the Cadence Design Systems System Custom IC and PCB (CPG) Group. He has over 22 years of experience in EDA technologies that enable the design and analysis of integrated circuit packaging and printed circuit board systems. Griffin is a graduate of Arizona State University.

Ron Huemoeller is Sr Vice President, Advanced Product Development, at Amkor Technology. Ron joined Amkor in 1995, and has served in several senior level roles with respect to Product Management & Development. Currently, Ron is responsible for corporate strategy, business development and deployment of advanced product platforms, including TSV, PLFO, Advanced Flip Chip, embedded die and MEMS devices. Prior to joining Amkor, Ron was Director of Engineering/Head Engineer at Cray Computer Corp. in Colorado Springs, leading the startup and development of state of the art motherboards for the world's fastest super-computer. Ron has authored numerous technical publications, co-authored the chapter on 'Assembly and Test Aspects of TSV Technology' in the Handbook of 3D Stacking (McGraw Hill) and has been granted more than 75 U.S. patents. Ron holds a B.S. in chemistry from Augsburg College with highest honors, a MBA in Business Management from Arizona State University and a Masters in Technology Management from the University of Phoenix.

Scott Jewler is a co-founder of SVXR, an innovative x-ray microscopy equipment company that designs and develops measurement and defect detection tools for advanced semiconductor interconnect including TSV, Si Interposers, fine pitch chip to chip bonding and other demanding applications. Prior to founding SVXR, Scott held executive positions at Amkor, STATS ChipPAC, and Powertech Technology, three of the six largest merchant semiconductor assembly and test design and manufacturing companies in the world. Scott led an R&D organization at PTI Taiwan that installed on of the first 3DIC manufacturing lines in the word with capabilities for TSV formation, wafer and interposer thinning, via expose, and thin die to die interconnect stacking. Prior to PTI, Scott led product organizations at both STTS and AMKR that directed more than a billion dollars a year each in semiconductor assembly and test business. Scott also has experience leading organizations responsible for package design, laminate substrate development, production planning, sales, and corporate strategy.

Larry Kinsman received a BS in Ceramic Engineering from Iowa State University in 1984 and an MBA from George Fox University in 2008. He began his career in the semiconductor industry at Inmos Corporation in 1985 working as a materials engineer on DRAM and SRAM packaging. He joined Micron Technology in 1986 working on hermetic packaging for high reliability and defense applications. At Micron he lead the package product development efforts for leads over chip, ball grid array, flip chip, and stacked wire bonded packaging for DRAM, SRAM, FLASH, graphics, and RISC microprocessor products. His focus changed to CMOS imager packaging design and product development in 2001 when Micron acquired Photobit. He was a Fellow at Micron from 1997-2008. He is currently a Distinguished Member, Technical Staff at Aptina Imaging where he leads the packaging design and development group. He currently holds 202 US patents.

Raj N. Master joined Microsoft in 2008. He is General Manager for IC Packaging, Silicon Operations, Quality and Reliability for all hardware products in Microsoft. These include Xbox, Kinect, Surface, Accessories, Zune, Keyboard, Mouse ,Webcam, and Roundtable etc. Raj was Corporate Fellow and Chief Technologistist for AMD from 1996 -2008. He was responsible to successfully transfer the IBM C4 / BGA technologies to AMD and set up high volume manufacturing in Penang which has to date produced more then 200 million flip chip assemblies. He led the Organic packaging development and manufacturing which is now in high volume production. As a part of that development he was responsible to select and develop package, component and material suppliers in USA to support high volume production. He is also responsible to qualify and provide technical direction to AMD bumping and probing operations in Dresden, Germany. He led the selection and qualification of Unitive and bumping foundry and Amkor and ASE as assembly and test foundries. He provides technical guidance for equipment and processes for C4 /BGA manufacturing lines in Suzhou, Penang and Singapore. He also provides technical expertise and guidance to product lines, Failure analysis, and reliability and quality organizations within AMD. He manages advanced packaging group involved in developing strategic enabling technologies .He is also manager of the Lead Free program of AMD. Raj joined AMD after spending 21

(continued)

Tuesday, September 24, 2013 • Biltmore Hotel • Santa Clara, California

2013 SEMICONDUCTOR ROADMAPS SYMPOSIUM

years at IBM. He was Senior Technical Staff member at IBM prior to joining AMD. He was responsible for packaging development and manufacturing as related to C4, Ball Grid Array, Column Grid Array, Board Level Reliability and Multi Layer Ceramic Substrate. Raj has 55 U.S. patents issued to him and has published over 80 technical papers.

Linda Matthew, Senior Analyst at TechSearch, earned her B.S. and M.S. degrees in Materials Science and Engineering from MIT. She was a development engineer at the IBM Watson Research Center in New York for seven years, working on leading edge flex packaging technologies. She joined Tessera when it had 19 employees and was based in New York, and subsequently moved with the company to Silicon Valley. At Tessera, she was responsible first for development, and later for technical marketing, of the µBGA®. She subsequently worked at nCHIP, doing technical marketing of MCMs, then at LSI Logic. She has numerous publications, and holds seven U.S. patents and four foreign patents. She has served as president of the Santa Clara, California IEEE CPMT chapter, and served for four years on the committee of the International Electronics Manufacturing Technology symposium.

Kenneth Potts is Cadence Design Systems Group Director of Strategic Marketing for Strategic Planning, and is responsible for leading the company's efforts in capturing and analyzing the data associated with all facets of the market and business landscape. He is also responsible for creating and managing all of the corporate planning processes. Prior to re-joining Cadence Design Systems in October 2010, Potts served as Virage Logic's Vice President of Asia and Foundry Sales responsible for leading the company's sales efforts in the Asia region and with its strategic foundry partners. Previously Potts served as Vice President of Product Marketing where he was responsible for leading the marketing strategy for the breath of Virage Logic's physical IP product line. Prior to joining Virage Logic in April 2005, Potts served as Vice President of X-Architecture Marketing for Cadence Design Systems. In this role, he was responsible for leading the product specification, market introduction, and proliferation of the key EDA technology required to enable the IC industry to design with diagonal wires. Potts holds a BSEE from Montana State University

Herb Reiter founded eda2asic Consulting, Inc. in the spring of 2002, after more than 20 years in technical- and business roles at semiconductor- and EDA companies, with the intention to focus on increasing the cooperation between EDA suppliers and semiconductor vendors. In this role he introduced innovative IC design tools to the major semiconductor vendors worldwide for several years. He also promoted the benefits of Silicon-on-Insulator technology, especially the transition from partially- to fully depleted SOI. As chair of the GSA's 3D-IC Working Group from early 2008 until end of 2011 and as SEMATECH business development consultant during 2012, he expanded his horizon further, to include 3D-ICs, packaging technology, semiconductor materials as well as manufacturing- and test equipment. Since January 2013 Herb also works on IC design- and manufacturing topics to enhance FinFET yields as well as accelerate failure analysis and production ramp-up of this important IC technology. Herb earned an MBA at San Jose State University and Master Degrees in Business and Electrical Engineering at the University and the Technical College in Linz/Austria, respectively. Herb took also more than forty Continuing Education courses at Stanford University in recent years.

Dongkai Shangguan, Ph. D. is currently the Founding CEO of the National Center for Advanced Packaging (NCAP China), in Wuxi, China, focusing on developing and commercializing advanced microelectronics packaging technologies for the industry. Concurrently, he also serves as an Executive Advisor and Laboratory Director at the Institute of Microelectronics, Chinese Academy of Sciences (IMECAS). Through his 20+ years with the industry, Dr. Shangguan worked at the Electronics Operations with Ford Motor Co. / Visteon Corporation in various technical and management functions, and at Flextronics as Corporate Vice President of Advanced Technology & Engineering Leadership. Dongkai received his BS degree in Mechanical Engineering from Tsinghua University, China, Ph.D. degree in Materials from the University of Oxford, U.K., and MBA degree from the San Jose State University. He conducted post-doctoral teaching and research at the University of Cambridge and The University of Alabama, and is currently a Guest Professor at several universities. Dr. Shangguan has published two books and authored/co-authored 250 technical papers and articles. He has over 20 patents issued. Dr. Shangguan is an IEEE Fellow, serves on the IEEE CPMT Society Board of Governors and is a Distinguished Lecturer for the IEEE CPMT Society. He also serves on the editorial/advisory board of several technical journals. Dongkai has received a number of recognitions for his contributions to the industry, including the "Outstanding Sustained Technical Contribution Award" from IEEE CPMT, "Leadership Award" from the Sustainable Electronics Manufacturing Working Group, "President's Award" from IPC, "Total Excellence in Electronics Manufacturing Award" from the Society of Manufacturing Engineers (SME), and the "Soldertec Lead-Free Soldering Award". He also received the "Distinguished MBA Alumnus Award" from the College of Business, San Jose State University.

Tuesday, September 24, 2013 • Biltmore Hotel • Santa Clara, California

 **2013 SEMICONDUCTOR ROADMAPS SYMPOSIUM**

Tom Strothmann manages Wafer Level Product Business Development for STATS ChipPAC, including Fan-in, Fan-out, and ultra-thin 3D Fan-out wafer level products. Prior to joining STATS ChipPAC, Tom was VP of New Business Development at FlipChip International. He has over 30 years of experience in semiconductor manufacturing, including extensive experience in wafer processing and flip chip wafer level packaging.

Dr. Hongxia Sun obtained her PhD degree in computer and electrical engineering and has worked in the industry for more than 14 years. Her current position is with Product and Technology Marketing team in STATSChipPAC in the semiconductor packaging industry, mainly focus on wafer level product strategy and management. She has previous experiences working in research labs and consulting companies as research engineer and technical consultant, with Experimental Laboratory for Internet Systems and Applications in Canada, Wireless Communications & High Speed Networks Laboratory in Michigan, and Telecommunication Research Center in Arizona. Dr. Sun has 34 publications in various conferences and journals. She was local organizing committee member for international conferences such as 16th International World Wide Web Conference, ACM SIGMETRICS'05, etc. and has been with several IEEE technical program committees since 2002, such as IEEE GLOBALCOM and ICC etc. She is also reviewer for international conferences, journals and academic publishers, such as John Wiley & Sons, Journal of Networks, International Journal of Communication Systems, ACM MobileHoc etc. Her expertise includes requirement engineering analysis, performance and statistical analysis, packaging modeling and simulation, market analysis and product management.

Anthony Torza
Biography not available at time of printing.

Paul Trio is the Senior Manager of Standards Operations at SEMI North America. He joined SEMI in 2002 as a Standards Engineer where he supported technical committees responsible for developing factory automation software, microlithography, and automated test equipment standards. He was graduated with a BS degree in Electrical Engineering from San Jose State University in 2001. In 2008, he received a dual Master's Degree in Systems Engineering and Business Administration from San Jose State. In addition to his responsibilities as Standards operations manager, he continues to support several technical committees including: environment, health, and safety (EHS); three-dimensional stacked integrated circuits (3DS-IC); high-brightness LEDs; as well as the North America Regional Standards Committee.

Suresh Upadhyayula has over 25 years of Management and Package Engineering experience in the valley in such diverse industries as component ASIC & memory packaging, SMT, LCD technology and packaging and HDI Substrate manufacturing. In addition he managed complex projects for new products introduction and Operations for relevant areas related to packaging such as quality & reliability, procurement and planning. Suresh has Masters' degrees in Chemical Engineering, Materials Science and MBA and attended the executive MBA program at Stanford.

Paul Werbaneth writes regularly for 3D InCites on 2.5D/3D IC technology and commercialization. Paul's other writing activities include his work as a guest editor for IEEE Transactions on Semiconductor Manufacturing, the contributed chapter on TSV Etching in the book "3D Integration for VLSI Systems," and an extensive number of articles, papers, and blogs regarding the semiconductor capital equipment business. Paul has worked as a business development manager at EV Group; vice president of marketing and applications at Tegal Corporation; country manager for Tegal Japan Inc.; senior plasma etch process engineer with Hitachi High Technologies; and as a hands-on process sustaining engineer in an Intel wafer fab. Mr. Werbaneth holds a Chemical Engineering degree from Cornell University, and recently completed studies in spoken Japanese in the Cornell Summer FALCON program.

Abe Yee is currently Sr. Director of Advanced Technology and Package Development at NVIDIA Corporation. His responsibilities include pathfinding and benchmarking technologies, investigating new technologies and setting NVIDIA's packaging roadmap for both GPU and Mobile products. From 2000-2002 he was Director of Engineering at SUN Microsystems responsible for SPARC processor manufacturing and reliability. He served as VP of Operations at Equator Technologies from 1996-2000 responsible for all aspects of development, NPI and production with manufacturing partners. From 1983-1996 he was at LSI Logic Corp, where he held various senior management roles in technology development and operations. Dr. Yee received his BA in Mathematics and Physics and his MA and PhD in Physics from UC Berkeley.

Tuesday, September 24, 2013 ∙ Biltmore Hotel ∙ Santa Clara, California

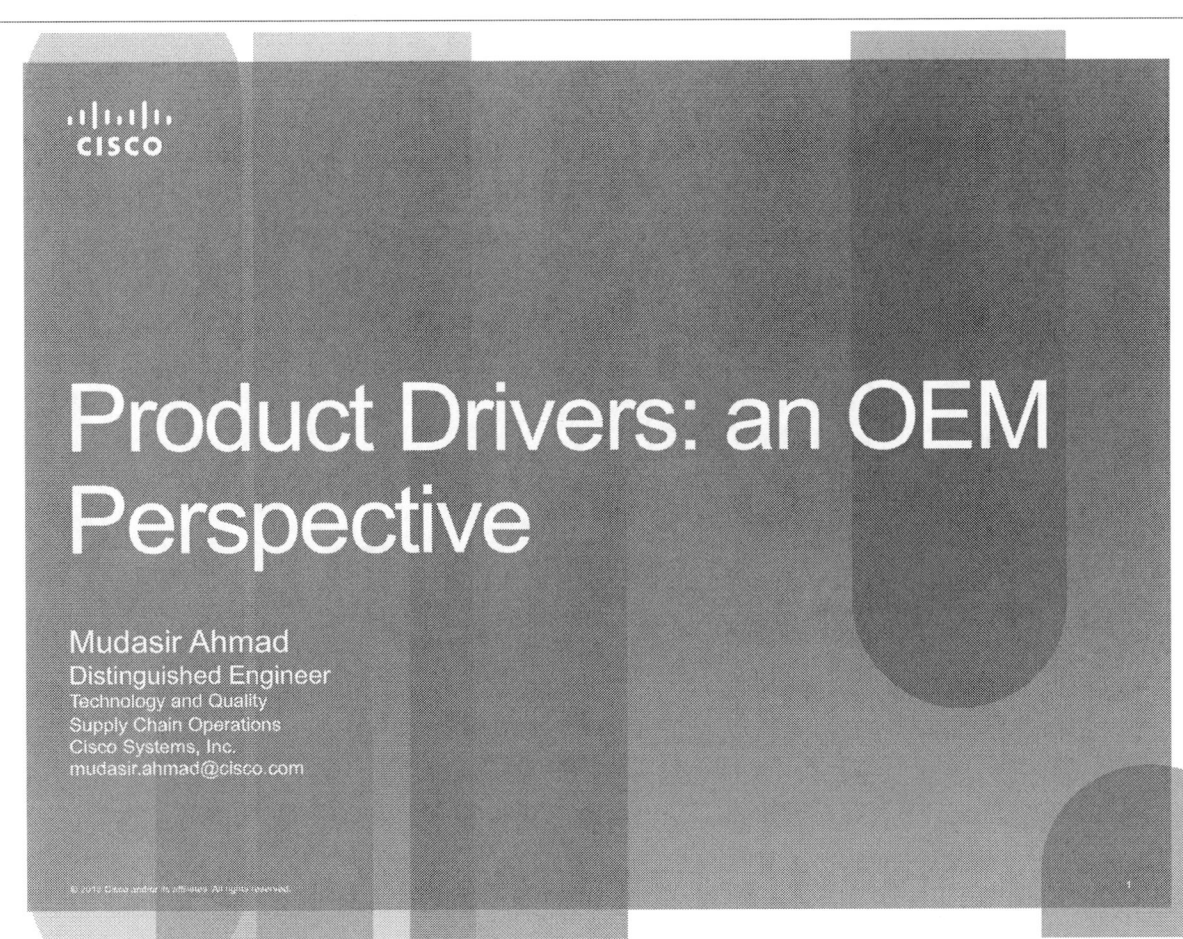

Product Drivers: an OEM Perspective

Mudasir Ahmad
Distinguished Engineer
Technology and Quality
Supply Chain Operations
Cisco Systems, Inc.
mudasir.ahmad@cisco.com

Major Disruptions Underway

- **Mobility**
  Globally, mobile data will increase 39 times from 2009 to 2014

- **Cloud and Virtualization**
  More than 25% of software likely purchased on a hosted basis over the next 5 years

- **Video**
  Within three years, 91% of all consumer internet traffic will be video

  Today, 50% of all Cisco IT Network traffic is video

http://www.slideshare.net/CiscoTurkey/cisco-services-portfolio

Cisco Product Portfolio Overview

Borderless Networks	Data Center Virtualization	Collaboration
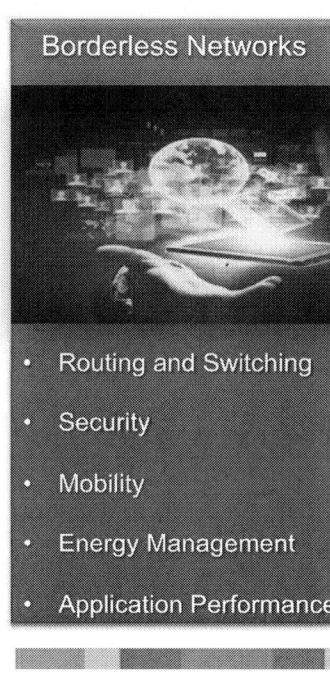	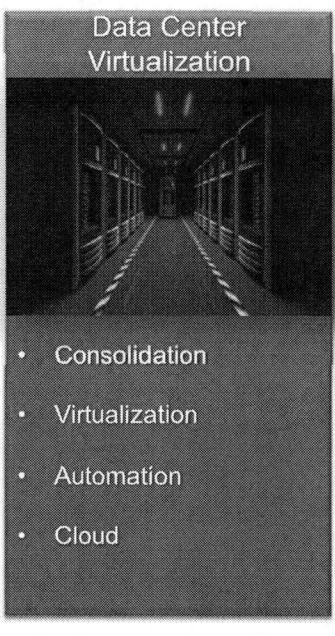	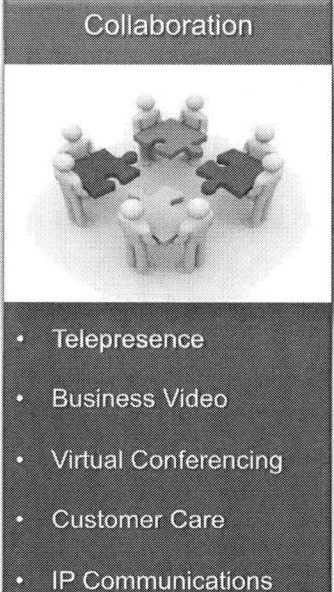
• Routing and Switching	• Consolidation	• Telepresence
• Security	• Virtualization	• Business Video
• Mobility	• Automation	• Virtual Conferencing
• Energy Management	• Cloud	• Customer Care
• Application Performance		• IP Communications

Fastest Core Router

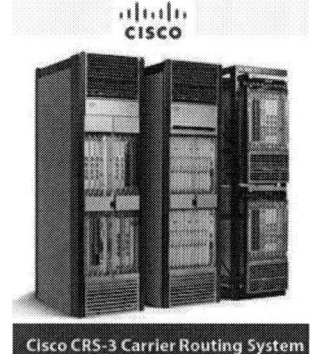

Cisco CRS-3 Carrier Routing System		
Cisco CRS-1 4-Slot Single-Shelf	4 slots	320 Gbps
Cisco CRS-1 8-Slot Single-Shelf	8 slots	640 Gbps
Cisco CRS-1 16-Slot Single-Shelf	16 slots	1.2 Tbps
Cisco CRS-1 Multishelf	1152 slots	92 Tbps
Cisco CRS-3 4-Slot Single-Shelf	4 slots	1.12 Tbps
Cisco CRS-3 8-Slot Single-Shelf	8 slots	2.24 Tbps
Cisco CRS-3 16-Slot Single-Shelf	16 slots	4.48 Tbps
Cisco CRS-3 Multishelf	1152 slots	322 Tbps

With 322Tbps, Cisco CRS-3 can...

... download the entire printed collection of the Library of Congress in just a second

... can deliver 1G to nearly every household in San Francisco

... enable every man, woman, and child in China to make a video call, simultaneously

... deliver all movies ever made in just 4 minutes

Product Drivers

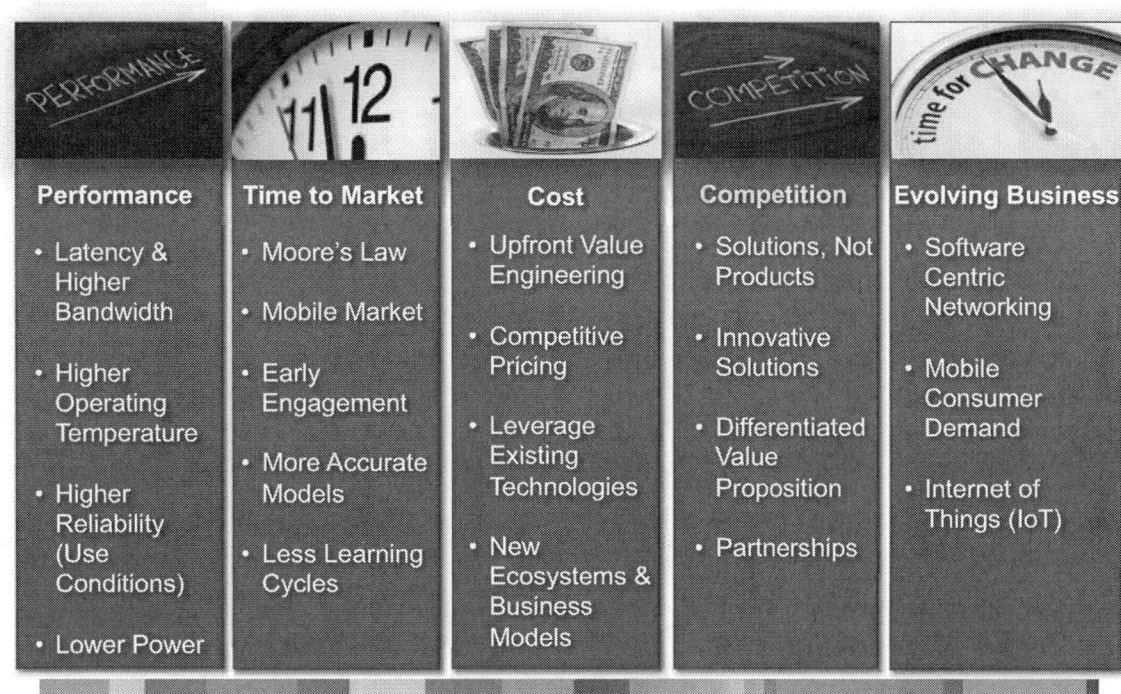

Performance	Time to Market	Cost	Competition	Evolving Business
• Latency & Higher Bandwidth	• Moore's Law	• Upfront Value Engineering	• Solutions, Not Products	• Software Centric Networking
• Higher Operating Temperature	• Mobile Market	• Competitive Pricing	• Innovative Solutions	• Mobile Consumer Demand
• Higher Reliability (Use Conditions)	• Early Engagement	• Leverage Existing Technologies	• Differentiated Value Proposition	• Internet of Things (IoT)
• Lower Power	• More Accurate Models	• New Ecosystems & Business Models	• Partnerships	
	• Less Learning Cycles			

High Performance Packaging Options

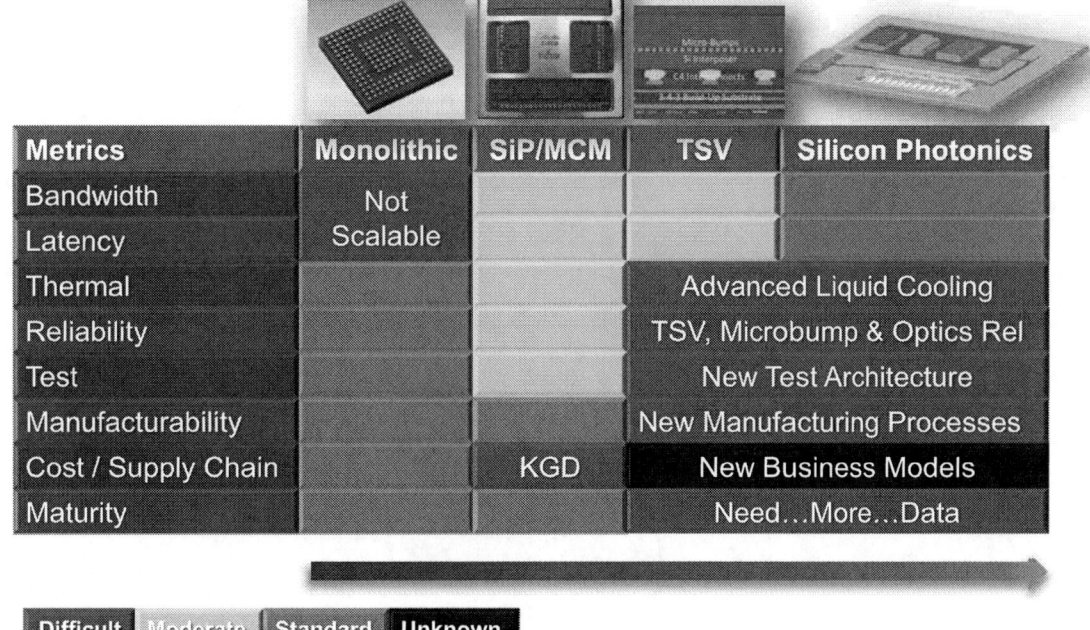

Metrics	Monolithic	SiP/MCM	TSV	Silicon Photonics
Bandwidth	Not Scalable			
Latency	Not Scalable			
Thermal			Advanced Liquid Cooling	
Reliability			TSV, Microbump & Optics Rel	
Test			New Test Architecture	
Manufacturability			New Manufacturing Processes	
Cost / Supply Chain		KGD	New Business Models	
Maturity			Need…More…Data	

Difficult | Moderate | Standard | Unknown

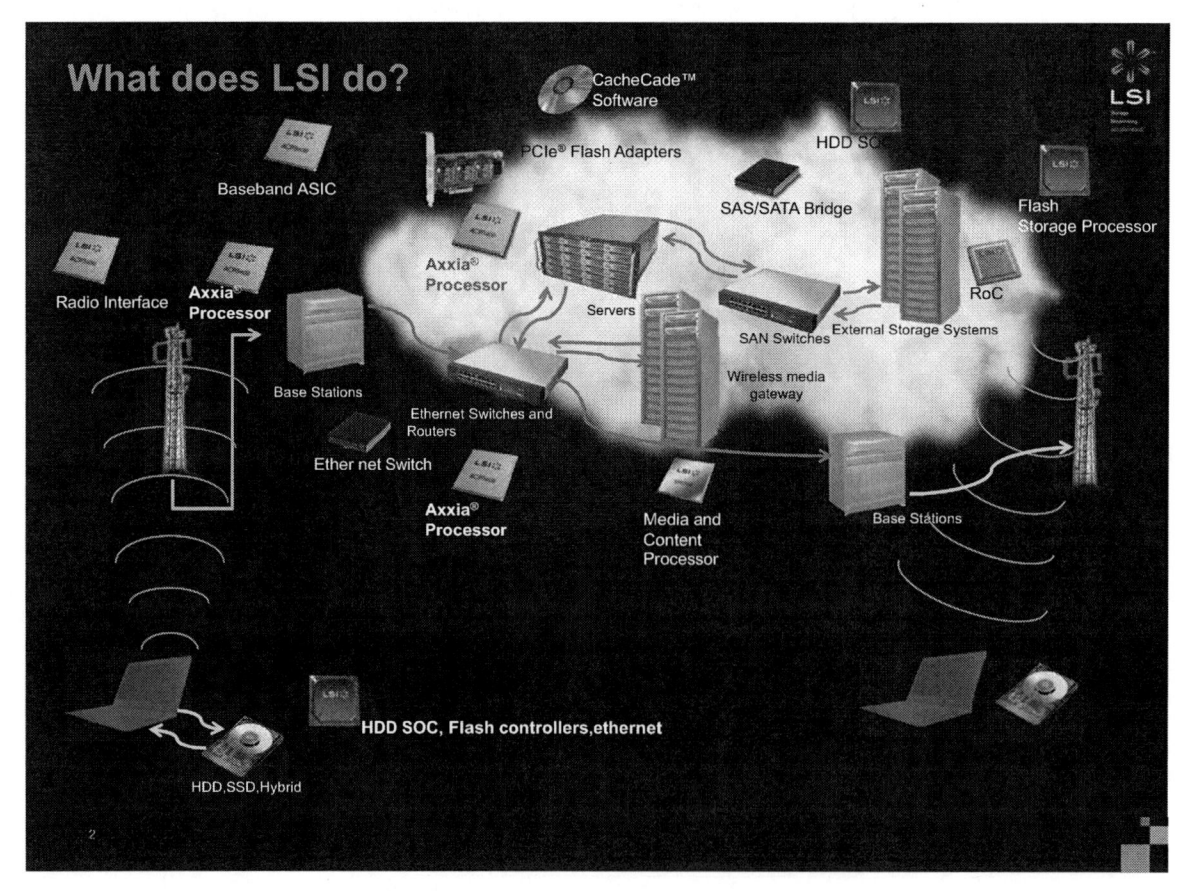

Product Drivers

- **Form Factor Geometry XYZ**
  - Shrinking on low end (xy <12mm) - while IO count increasing
    - Finer BGA pitch - System board co-design
  - Growing on high end (xy ≥ 55mm)
    - High IO count, die size
    - MCM
  - Aggressive z height - Storage (<1mm)

- **Performance with high data rate IO's**
  - New materials
  - Finer design rules
  - Increasing thermal requirements

- **System Integration**
  - Silicon / Package / PCB signal and power integrity
  - Board congestion and complexity
  - Manufacturability
  - Cost

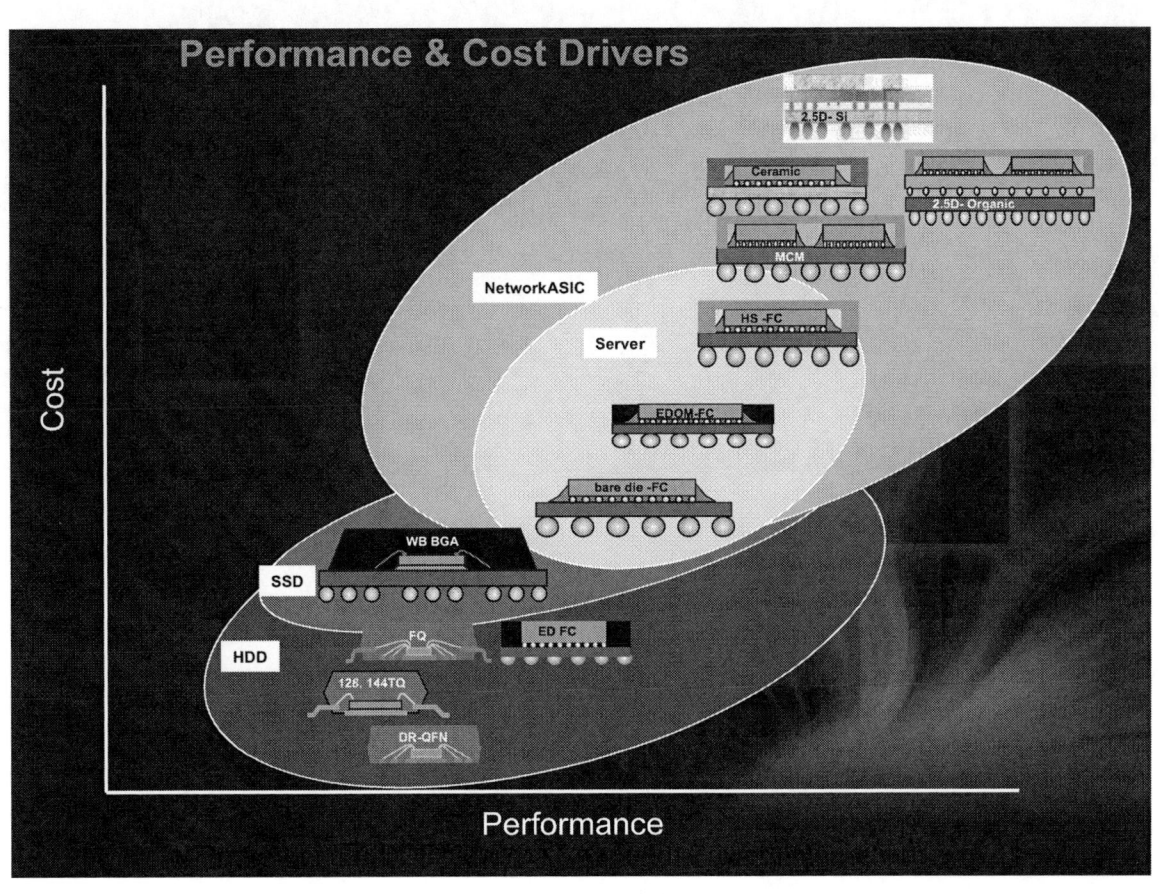

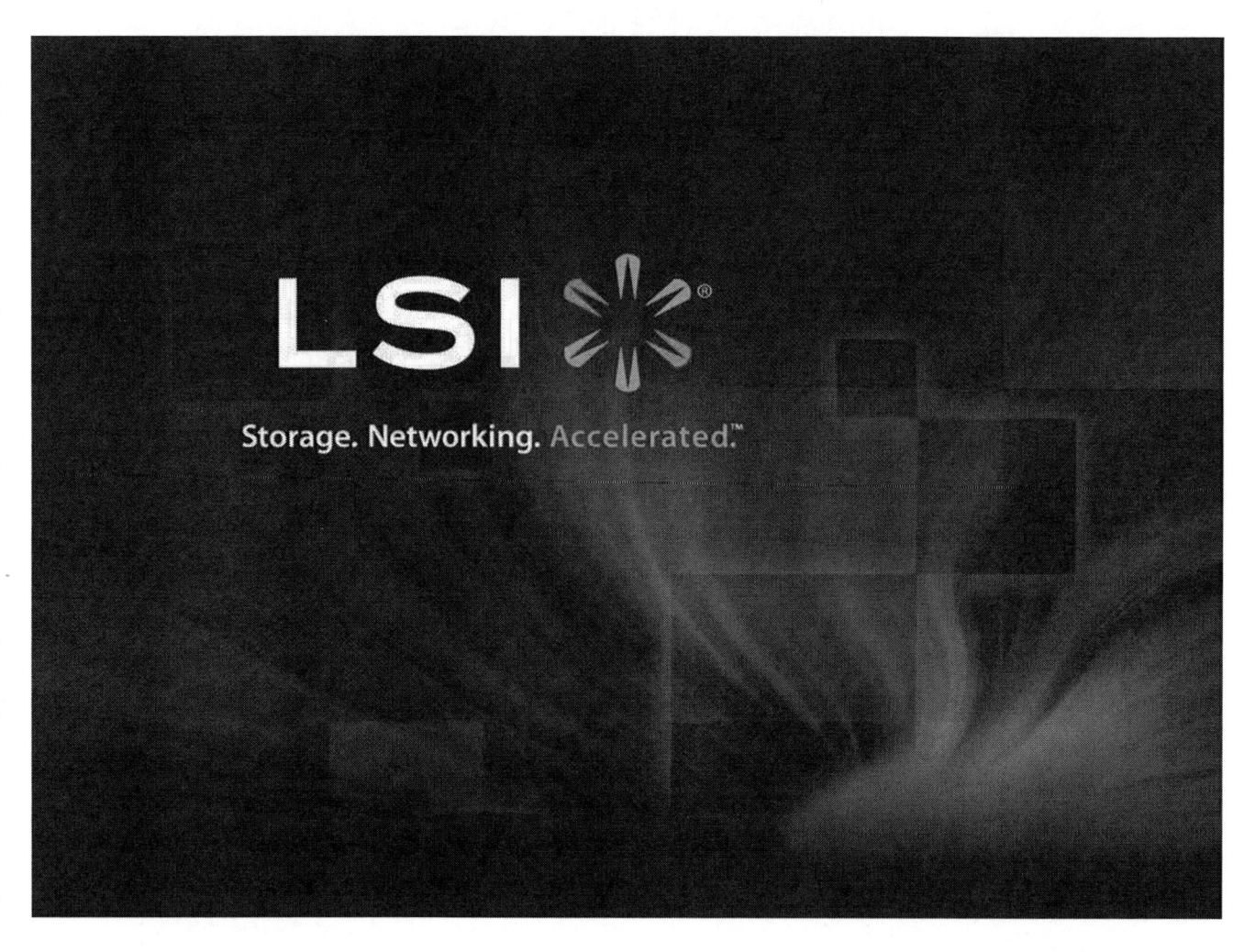

CMOS Imager Sensor (CIS) Packaging & Product Drivers

Larry Kinsman

Distinguished Member, Technical Staff

Package Product Development

September 24, 2013

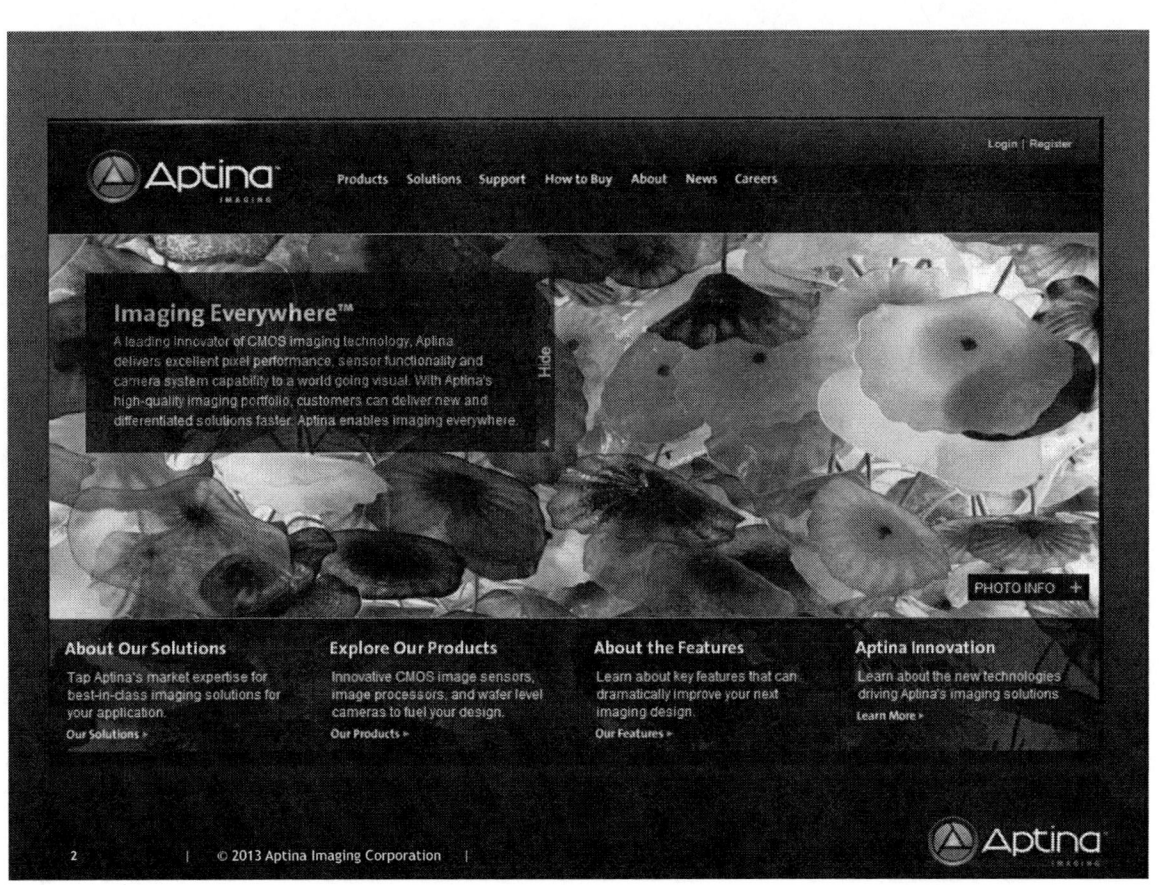

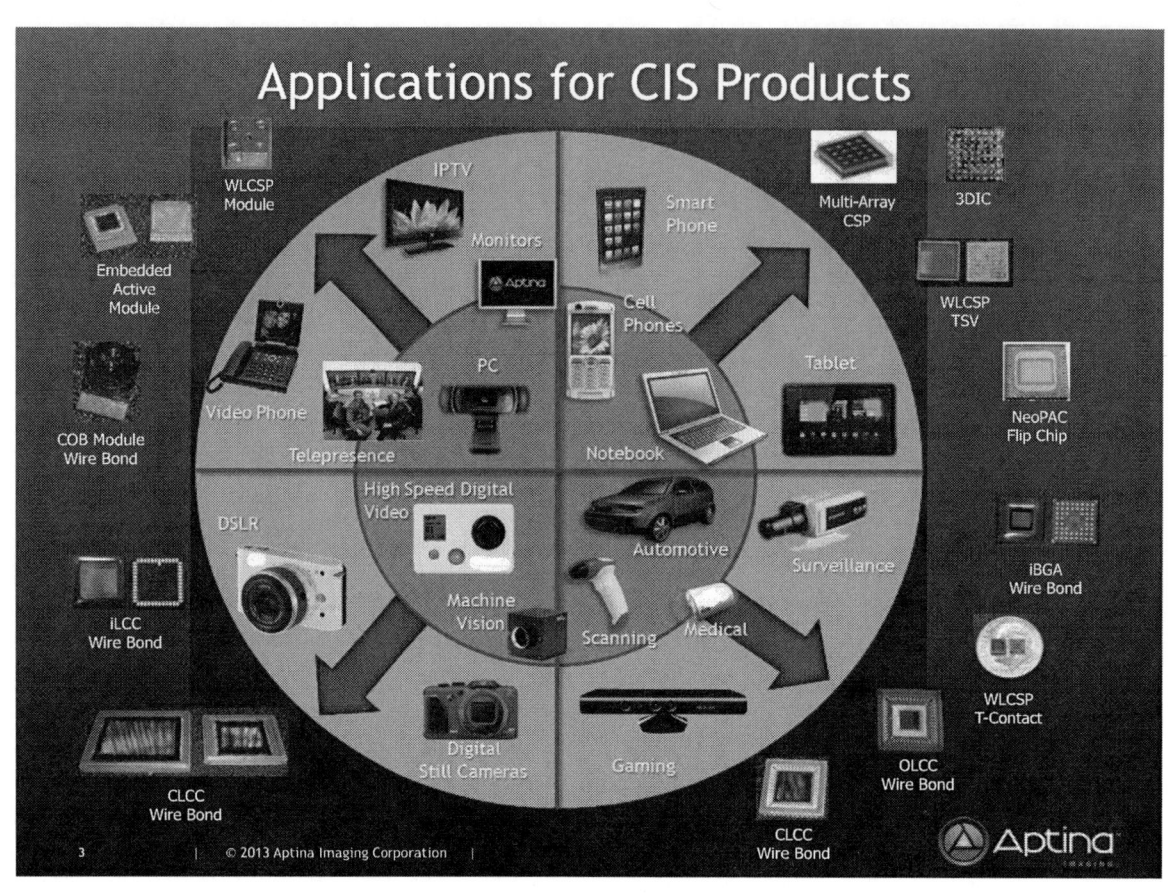

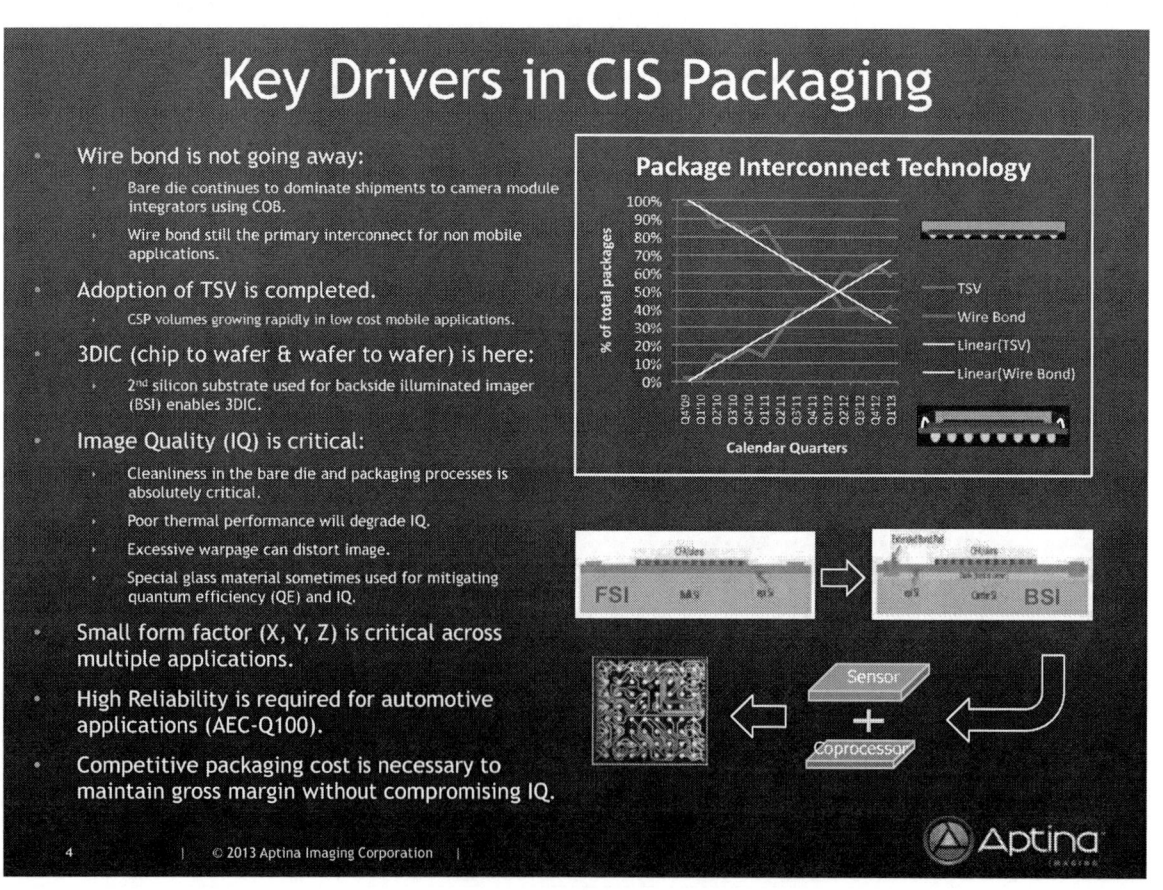

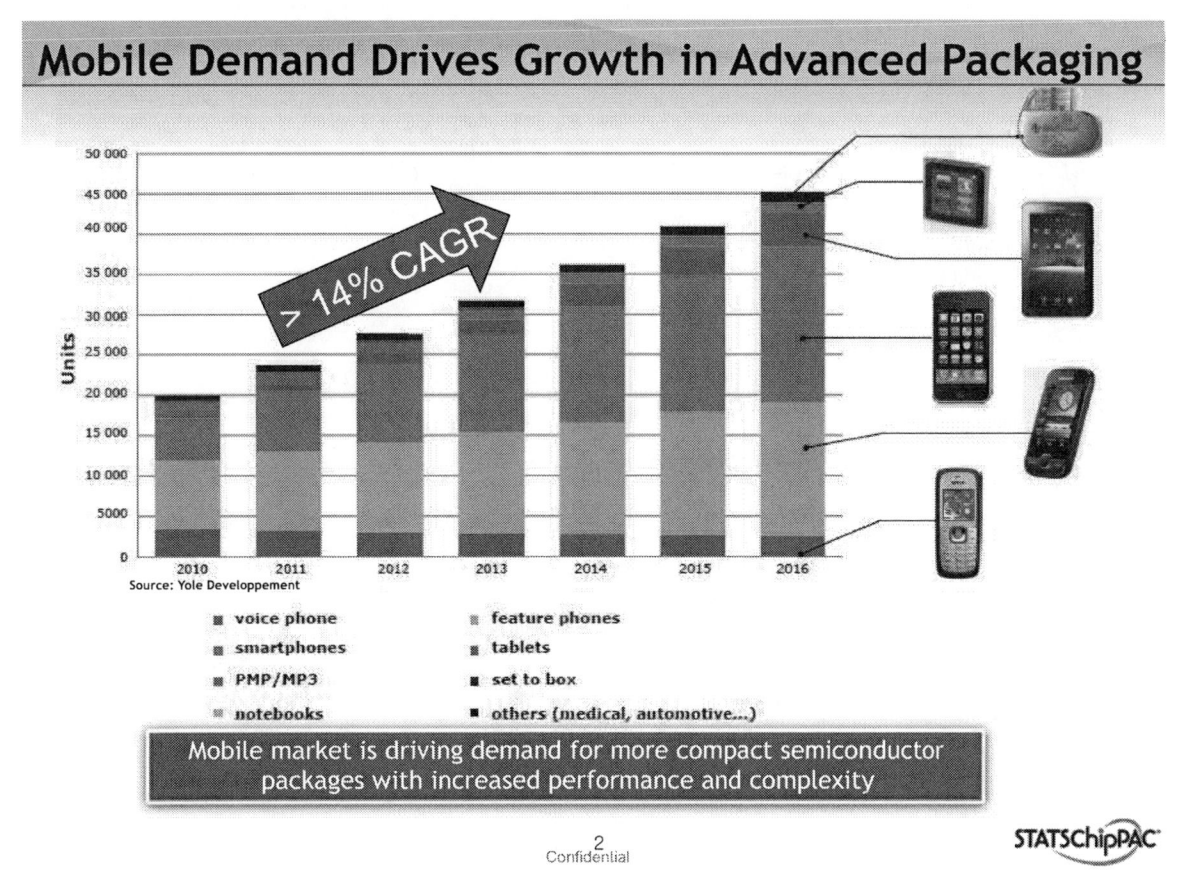

Next Generation Packaging Requirements for Mobile

- Shrinking silicon wafer nodes: 32nm > 28nm > 20nm
  - Wafer node reduction impacts assembly

- **Shrinking product form factor**
  - Ability to handle increasing I/O count per silicon area
  - Reduced package thickness to meet smaller form factor

- **PoP thickness reduction to < 1.0 mm including 2 die memory**
  - BGA thickness 0.8mm

- **System Integration – Enable easy platform design at customer level**
  - Integration of embedded active and passive components in PoP and SiP solutions to reduce the overall number of packages in a product
  - Package ball pitch in-line with customer PCB build-up targets

- **Thermal Performance – Peak performance limited by package design**
  - Non-POP side-by-side memory solutions for tablets
  - Exposed die heat spreader for application processors
  - New materials for increased thermal conductivity

- **Manufacturing and test solutions for 2.5D and 3D integration**
  - Backside processing, assembly and test of wide-IO application processors

- Cost - Cost - Cost
  - Cost reduction strategies for current and future technologies

STATSChipPAC

Driving PoP Height Reduction with eWLB

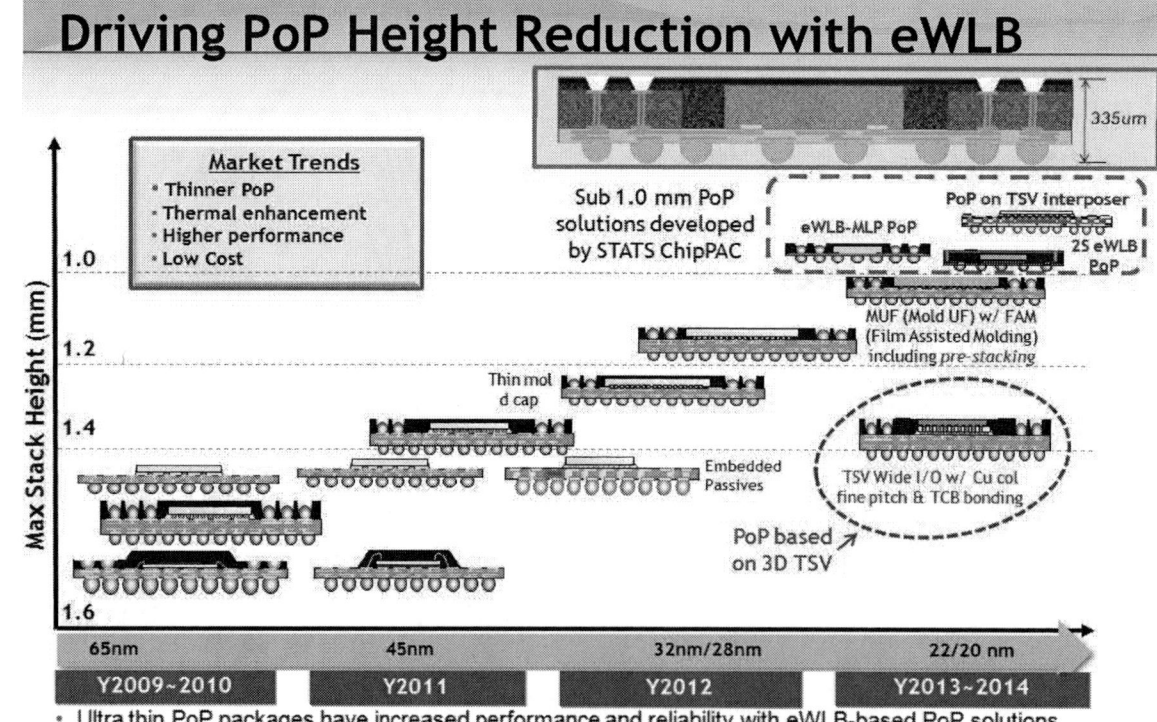

- Ultra thin PoP packages have increased performance and reliability with eWLB-based PoP solutions providing customers a competitive edge over substrate-based PoP technology
- eWLB PoP can provide >500 I/O in an overall thinner package using a dense vertical interconnection

STATSChipPAC

SanDisk®

Drivers & Packaging Challenges at SanDisk

Suresh Upadhyayula

Sr. Director, Pkg Eng Management

13-Sep-2013

SanDisk Products

Packaging Technology Challenges

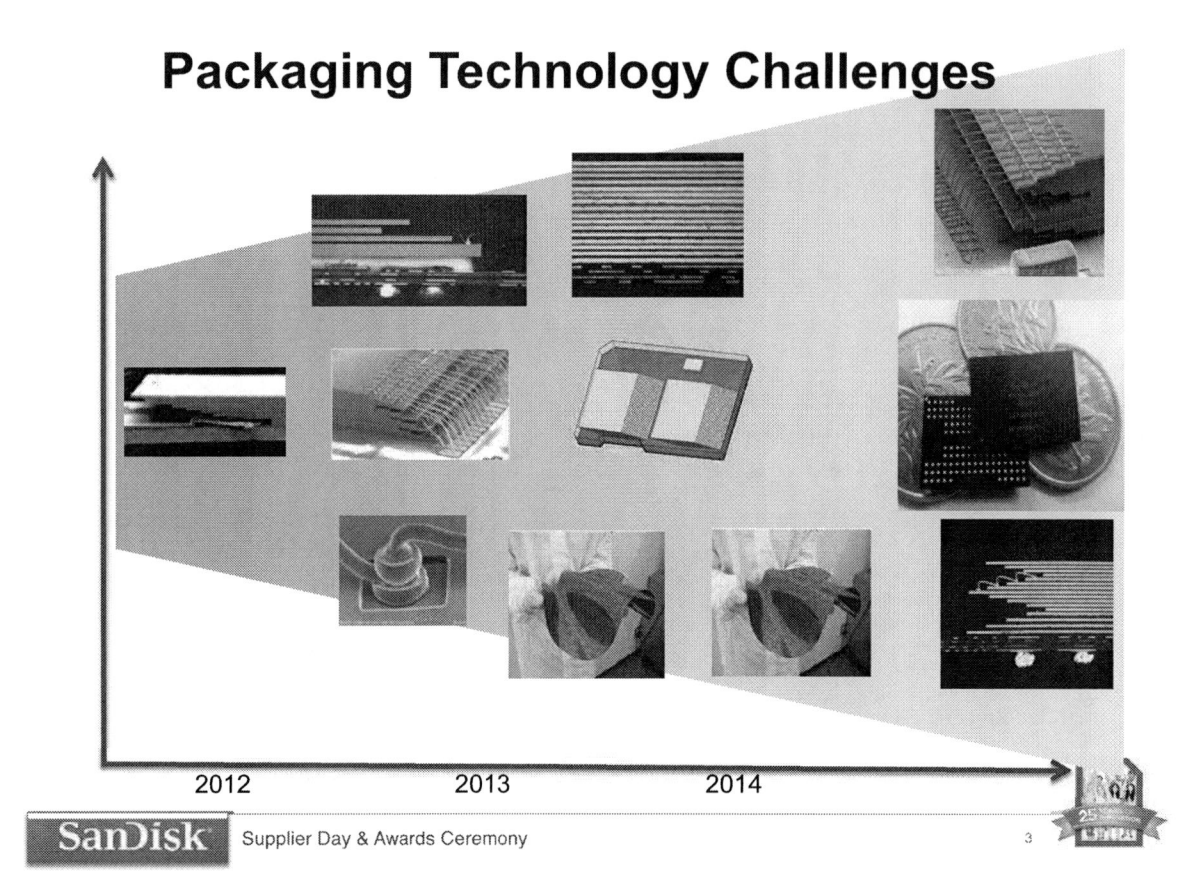

2012   2013   2014

SanDisk Supplier Day & Awards Ceremony

Drivers

❖ Cost

    ❖ 15%-30% cost reduction per GB QoQ

❖ Quality/Reliability

    ❖ Semi-Conductor Quality / Reliability Standards

❖ Capacity

    ❖ Stacking more and more die: 8D → 16D → 32D

❖ Form-factor

    ❖ Die smaller than Package by <100um on each side

❖ Cycle Time (Time to market)

    ❖ Wafer in to finished product out in 24hrs

SanDisk

MEPTEC Roadmaps 2013
Session 1 Notes

Paul Werbaneth

3D InCites

www.3dincites.com

Session 1: Product Drivers

Panel Moderator:

Joel Camarda, Amonix

Panelists:

Mudasir Ahmad, Cisco

Farshad Ghahghahi, LSI

Larry Kinsman, Aptina

Tom Strothmann, STATSChipPAC

Suresh Upadhyayula, Sandisk

Panel 1 Charter

Product requirements drive both wafer process nodes and package technology forward. Processor, ASIC, and FPGA advances have driven the industry to nano-scale geometries and 300 mm wafer size for silicon, and to flip chip interconnect and 2.5 / 3D technologies for packaging. Requirements in device functionality, power, speed, and mobility continue to drive the packaging roadmap. **We shall assemble a panel of product packaging experts**, including representatives from different semiconductor device and system markets, **to discuss their packaging roadmaps and the challenges therein**.

Mudasir Ahmad, Cisco

- Not a morning person
- Cisco looking at disruptions in the network for new business opportunities – smartphones, cloud, video; projecting **91% of future internet traffic will be video** – Netflix drives internet traffic today, Cisco loves this because they build routers that drive internet traffic
- Product driver targets example: **10X performance @ ½ cost @ 5 times reliability**
- Operating temperature opportunities – costs as much to cool products as it does to run them, want higher temp operation w/ reliability to save cost – desert data center
- Metcalfe's Law – IoT grows with the square of the nodes, not just Moore's Law growth
- Cisco maybe not a large customer compared to mobile – sometimes hard to get memory supplier's attention
- Cisco now into SiP/MCM 14 memories and ASIC – **end game is not just TSV / 3D IC – it is silicon photonics** ... but need new business models and things like advanced liquid cooling
- How to test, model, define failure modes for new technologies – drives qualification requirements
- **Reliability becoming segmented**

Farshad Ghahghahi, LSI

- Pure play semiconductor company focusing on data transfer and data storage using ASIC and SoC
- Product driver challenges: form factor geometry – finer BGA pitch, increasing I/O, **decreasing height <1mm**
- Need new materials and other ways to **better accommodate thermal handling** – 100W+ needs to be dissipated
- Need to pay attention to system integration – silicon/package/PCB
- For packaging, HDD and SSD applications relying on flip chip– server market also flip chip of various flavors – network market ASIC MCM
- Looking at 2.5D and **silicon photonics**

Larry Kinsman, Aptina

- 202 patents!
- Aptina is CMOS imaging technology company – imaging chips and algorithms **market forecast 17% CAGR**. Imaging Everywhere ™
- Small format phone sensor up to Nikon DSLR, automobile
- WLCSP w/ TSV, embedded active module (includes passives); **large portion of business is shipping bare die to module integrators**
- New requirement is to handle optical function along with everything else
- Wire bond still prevalent, TSV introduced and made product crossover over a year ago – **at the forefront of adopting TSV.**
- iPhone has Sony stacked chip solution sensor to coprocessor **wafer-to-wafer 3D interconnects** – transition because of backside illumination – **can be quite economical** – reduced chip size
- Thermal performance interacts with dark current which is read as image noise
- Die flatness, warpage can distort image
- High reliability for automotive applications, challenging for small form factors, need air cavity above the sensor array – open cavity not hermetic package
- **Drive performance up, but it can't cost any more**, can't sell parts based on performance alone.

Tom Strothmann, STATSChipPAC

- Trained a lot of people how to do bumping
- Focus today on wafer level products – Yole says **>14% CAGR for mobile drives growth in advanced packaging** – WLCSP, 3D IC, Fan-out products (allows extending the 6x6 chip limit to 14x14 and above) also provides flexibility for 3D
- Shrinking wafer nodes impacts assembly, .4mm pitch to .35mm pitch for I/O 980 balls at .4 pitch on 14x14 package
- Want PoP thickness <1.0mm including 2 memory die
- Incorporate more components, chips, passives which can be done with fan-out products
- **Thermal performance**
- Manufacturing and test for 2.5D and 3D integration – challenging with TSV structures
- **Cost, cost, cost** – mobile market demands this, can't develop exotic solutions without considering cost
- **Reliability – for mobile, lower than some other markets**, low-end suppliers in China for China market driving this e.g. <500 thermal cycles to reduce cost
- Driving PoP height to under 1mm application processor with stacked memory <1mm high of great interest to the mobile guys
- Process / product development in Singapore, including TSV

Suresh Upadhyayula, Sandisk

- Let the pictures speak for themselves for SanDisk products
- Packaging challenges: stacking die from 8 to 16 to 32 (target), possibly some thermal challenges, **wafer handling challenges for 25um thick wafer**
- **15%-30% cost reduction per GB QoQ** is a driver means SanDisk costs need to come down the same way
- Quality / Reliability/ Cost balance - shifting from enterprise to mobile
- Form factor – die smaller than package by <100um per side – maybe going to 50um per side
- Cycle time – time to market – San Disk reduced **wafer in to finished product to 24 hours (!)**

Audience Questions (1)

JC: Comments about reliability target in the mobile market, how does it influence engineering, etc.?

- MA: **Segmented vendor list** and where product will go in order to know where reliability may go. Server and router have different reliability requirement – al a carte negotiated pricing based on segmented reliability – profit margin gets baked in. Not too easy to do the segmentation. This is hard, Cisco fabless, and solution seller with OEM selling to Cisco – need to work with OEMS. **"High touch" environment**. Negotiation. Q-Q reductions set up at the beginning.
- SU: Similar product segmentation – retail market reliability less than enterprise market requirement. Big challenge, but even enterprise market cost sensitive.
- LK: Focusing on Tier 1 and 2 mobile – **China market not demanding same reliability**. Automotive market is different – need better reliability , but still has cost pressure. There are conflicts.
- TS: Mainstream mobile; see on the horizon that **reliability requirements will drop, particularly in China**. Auto market will not shift from the 1000 cycle requirement – mobile is 500 cycle, China may be 300.
- FG: Two markets: in the enterprise market reliability requirements are key.

Audience Questions (2)

Audience: For Tom, looking at reliability, **TCE is different for different materials**, mechanical stress. Comments – what are you doing?

- TS: Need to deal with the constraints brought on by the materials, maybe by tailoring mold compound TCE to be like silicon. Historically, WLP non-underfill application, but it's becoming more common to do underfill, which helps. **Tailor the materials**.
- LK: Glass, silicon chip, organic substrate – need the image to stay flat over a certain temperature range, but the image sensor changes shape over that range – challenge.

Audience: Product driver network speed increases thermal load on system, suggests you want to be tolerant of higher operating temperature. **What kind of junction temperature do you want to see?**

- MA: **As high as possible**. 105C, but need to look at temperature for silicon photonics, which is limited to 70C for performance reasons. Also, memory limit is 85C, now pushing memory suppliers to 95C and above junction temperature.

Audience: Silicon photonics is becoming important – as soon as you fix one problem another one comes up. What is the gorilla-in-the-room bottleneck?

- MA: The gorilla is the supply chain and it's the TSV supply chain. We love TSVs, but **the supply chain hasn't yet taken on the responsibility for reliability**. TSMC has been doing good things. Want organic interposer; silicon is good from TSC standpoint. Big challenge is still the business model.

Audience Questions (3)

Audience: We've heard a lot about silicon interposers, and now organic interposers. Comments about glass as the solution for an interposer?

- (*Stage direction: all panelists look at each other*) MA: **Glass is a very happy medium**; if we can get the pitch density we need we're interested in glass, but it's not there right now compared to silicon. Were it to be, we would be interested. Why organic? Broad array of low-cost interposer suppliers and good supply chain with organic interposer.
- FG: More and more the supply base is offering products where the line density is becoming more feasible. **Silicon has physical size limits which organic does not** – also organic supports multiple vendors – supply chain advantage.
- TS: **Glass is interesting but the technology is immature. You can't beat silicon** for line width capability. Other technologies will be coming on line for small package sizes with fan out.

Audience Questions (4)

Audience: Aptina has been active in TSV for some time, The memory guys are coming on now, but the TSV roadmap keeps getting pushed out. Beyond optical sensors and memory **what's next for TSV**?

- SU: Looking at TSV for flash, but TSV need is for high bandwidth high I/O – flash is not there yet.
- MA: We're there, and we've been there for a while, but the forecast don't add up. **Better to push silicon monolithic as long as possible than invest in a new TSV ecosystem.** Need higher performance at lower cost; hard to deal with the investment / cost required for TSV. Like the air cooling – liquid cooling example.
- TS: TSV is there now – we have the capability, it's a matter of cost – everybody struggles with competing against equivalent performance w/o TSV. **TSV will come; it's just a matter of time and the economics.**

Panel Two: Manufacturing Drivers in Semiconductor Roadmaps

Moderator: Jeffrey C. Demmin, STATSChipPAC

Panelists:

- Richard Crisp - Invensas Corporation
- Javier DeLaCruz - eSilicon Corporation
- Ron Huemoller - Amkor Technology, Inc.
- Raj Master - Microsoft
- Dongkai Shangguan, Ph.D. – NCAP China

2013 MEPTEC
SEMICONDUCTOR ROADMAPS SYMPOSIUM

Session 2: Manufacturing Drivers in Semiconductor Roadmaps

It is more important than ever that companies consider their supply chain partners -- both upstream and downstream -- when creating their roadmaps for their own businesses. For example, much of the value of semiconductors is in miniaturization and performance advances, which is a function of packaging but of most value to the OEM. A company in any of these three segments -- OSAT, IC, and OEM -- must make sure that its plans for the future are integrated with those of other companies throughout the supply chain.

This panel discussion will allow representatives of all of these types of companies to discuss how they interact with suppliers, customers, and their customers' customers to help their businesses move forward with coordinated roadmaps.

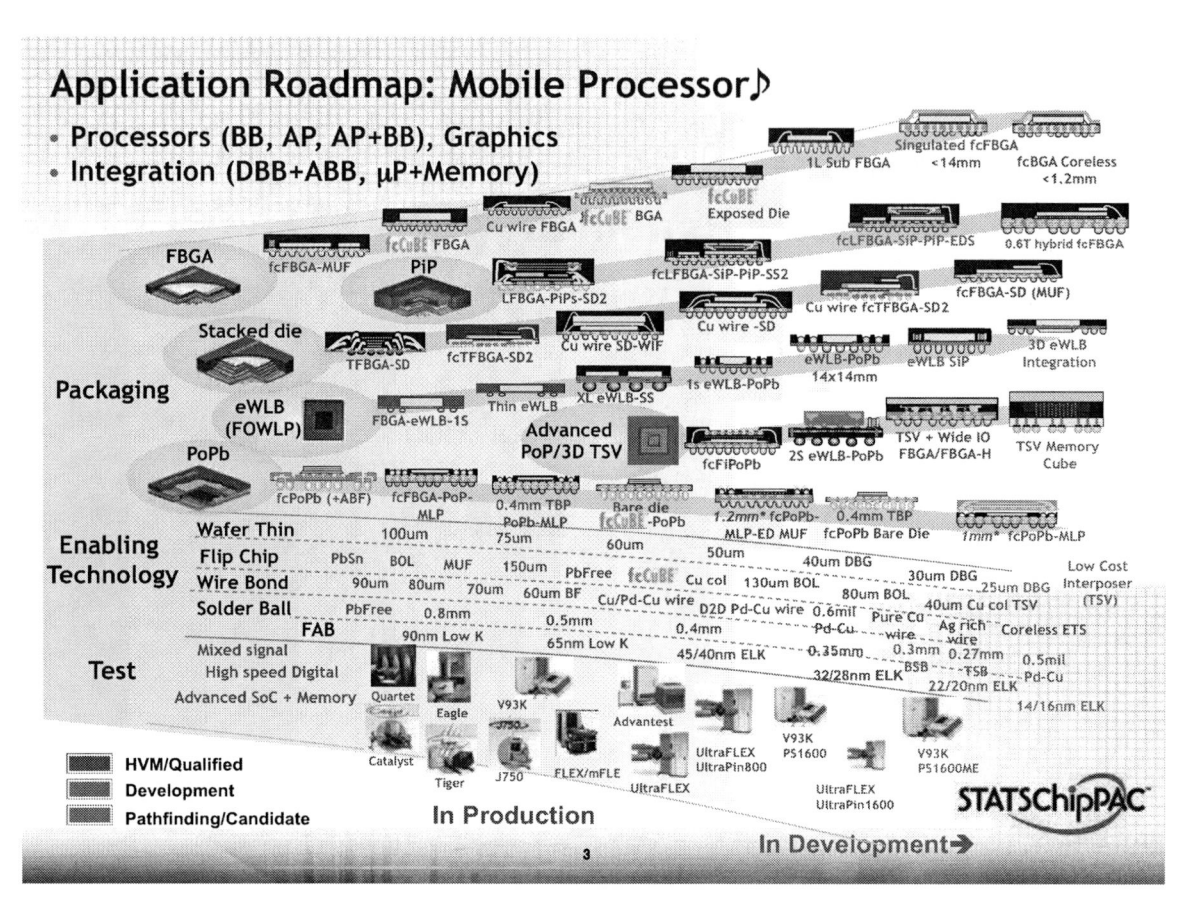

Richard Crisp

Vice President and Chief Technologist, Invensas Corp.

rcrisp@invensas.com

September 2013

- **Formation:** Founded in 2011 as a wholly owned subsidiary of Tessera Technologies, Inc. (Nasdaq:TSRA).

- **Goal:** Develop and commercialize breakthrough semiconductor interconnect solutions and IP in mobile, storage and consumer electronics.

- **Core Focus:** "Interconnectology": advanced interconnect, semiconductor packaging, memory circuitry, modules, 3D TSV architecture.

- **Company:** 50+ Employees (1/3 PhD). **Headquarters:** San Jose, CA.

- **IP:** >1000 patents and applications.

Major Trends Driving Memory Packaging

- Notebook PCs
  - 4GB-8GB memory (DDR3L and LPDDR3)
  - Solder-down to motherboard (Ultrabooks)
  - HDI PCBs for smallest motherboard size (battery size tradeoff)
  - Moving away from SODIMM
  - Opportunity for multi-die DRAM packaging is growing
- Smartphone Handsets

  - Focus on performance and power reduction
  - POP generally preferred for AP to Memory connections
  - Memory bandwidth moving above 25GB/sec in 2014, up to 100GB/sec in three years from 2-4GB of LPDDR3 or other type DRAM

  - Key challenges: cost, risk, power, size
  - Opportunity for improved POP vs TSV for near term
- Servers
  - Focus on density, cost and power
  - Frequently requires DDPs for density

Notebook Trends

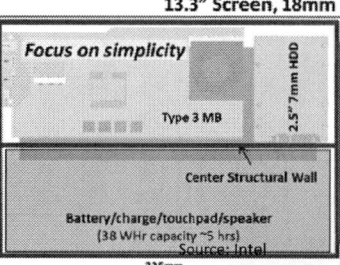
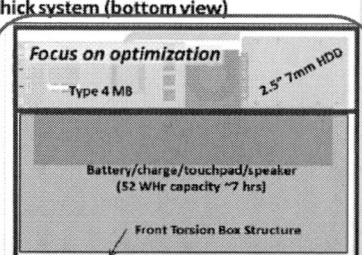

Low cost
But big PCB

Reduces
battery
capacity

High cost
But small
PCB

Increases
battery
capacity

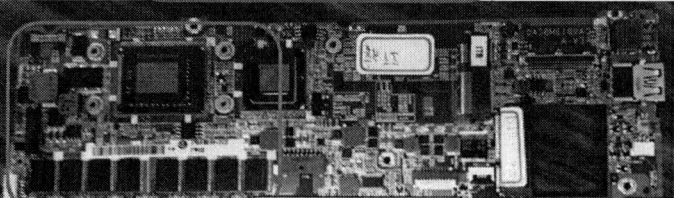
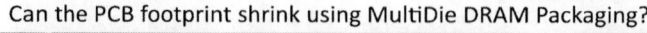

Can the PCB footprint shrink using MultiDie DRAM Packaging?

Invensas DIMM-in-a-Package

PoP Trends

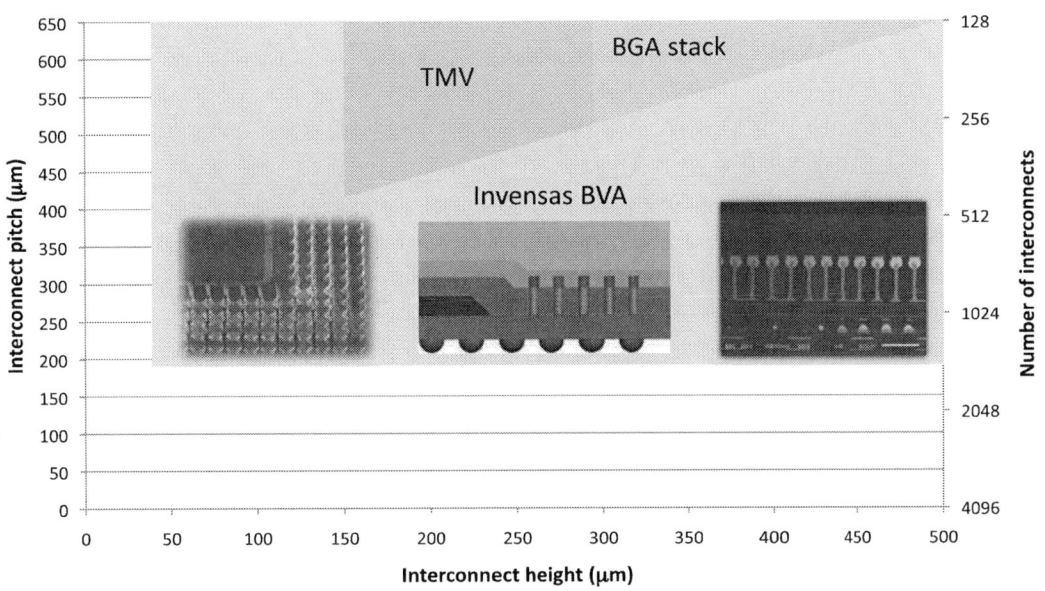

- BVA substantially improves the pitch and number of IO while being independent of the interconnect height (distance between logic and memory substrates)

Server Trends

- DIMM oriented
- Capacities today:
  - RDIMM/LRDIMM: 32GByte (4Gbit die with DDPs)
  - VLPRDIMM: 16GByte (4 Gbit die with DDPs)
- DDP challenges
  - High cost
    - Structure uses RDL
    - Complex manufacturing flow
  - Low Performance
    - Long in-package stubs
  - Thermal
    - Die stacked on top of each other

Conventional RDL DDP

Invensas xFD DDP

Backup

Dell XPS-12 Ultrabook: System PCB Layout

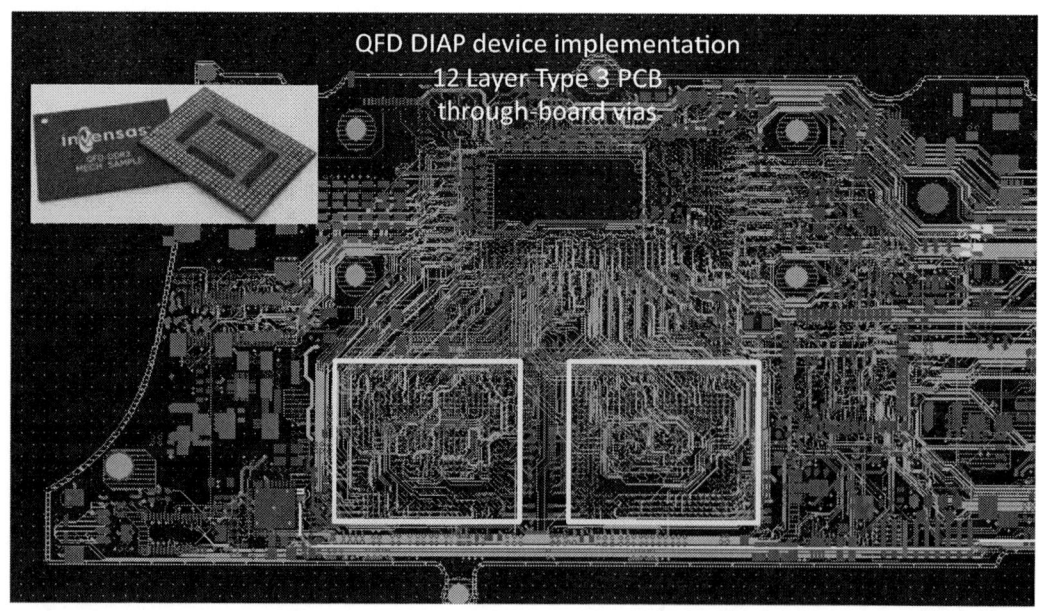

BVA Formation: wirebonding & molding

14 x 14mm SOC, 1020 leads; 5 rows at 240 micron pitch

K&S ICONN Bonder

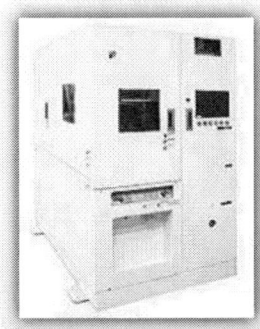

Yamada G-Line Mold Machine

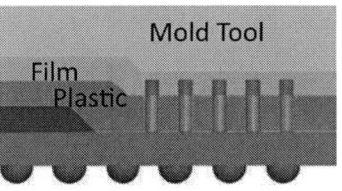

Film Assist Molding

Mold Tool

Film
Plastic

BVA PoP Stacking: Results

UIC Advantis3 SMT

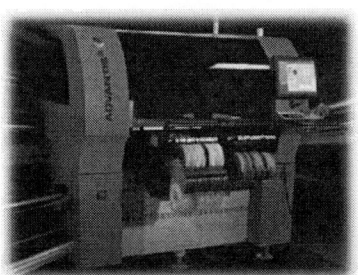

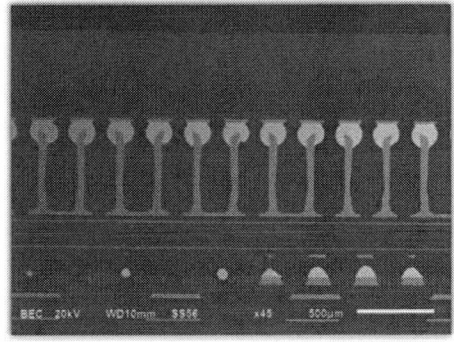

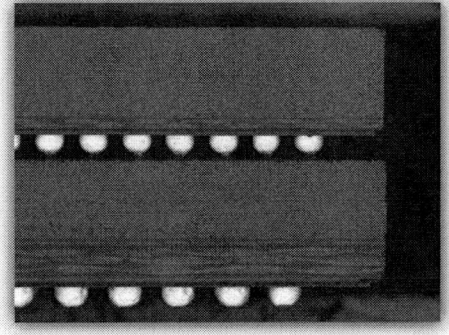

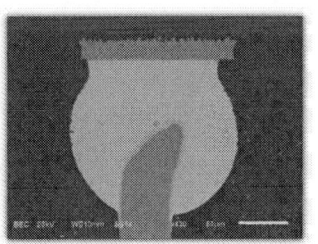

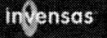

Invensas Dual Face Down (DFD)

Invensas Dual Face Down

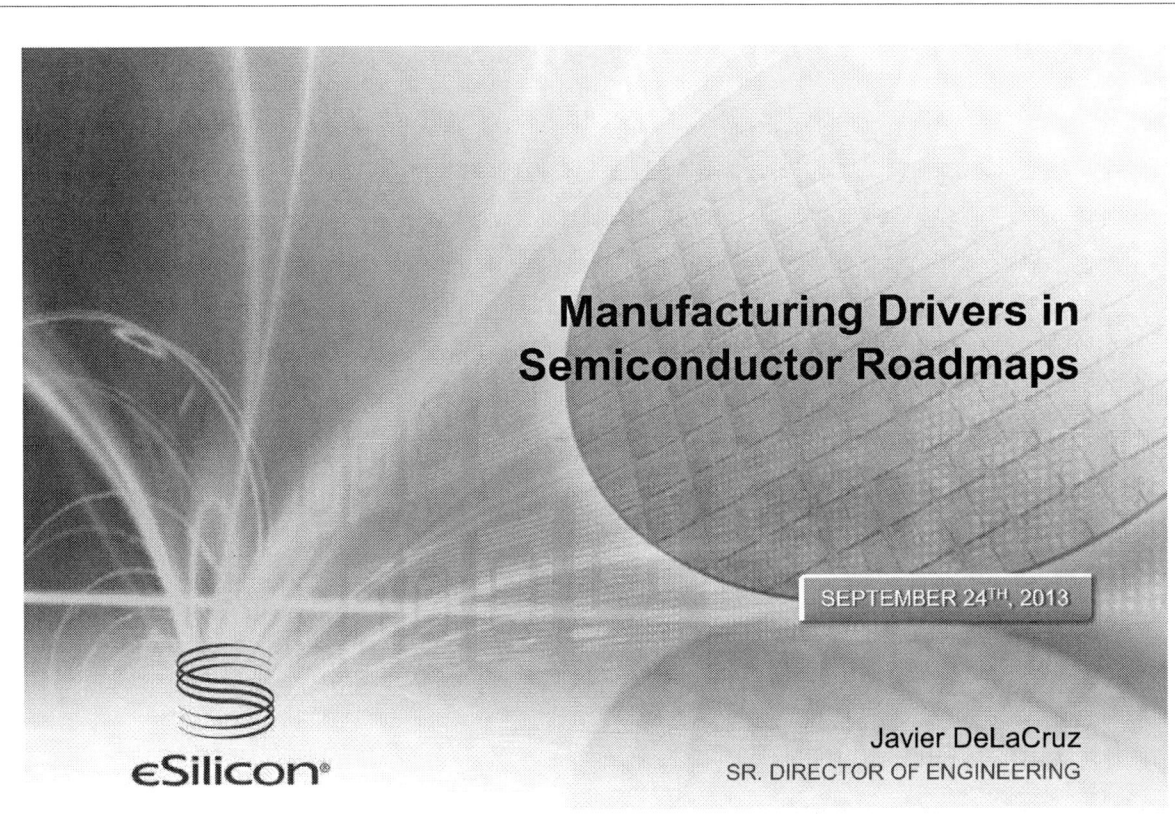

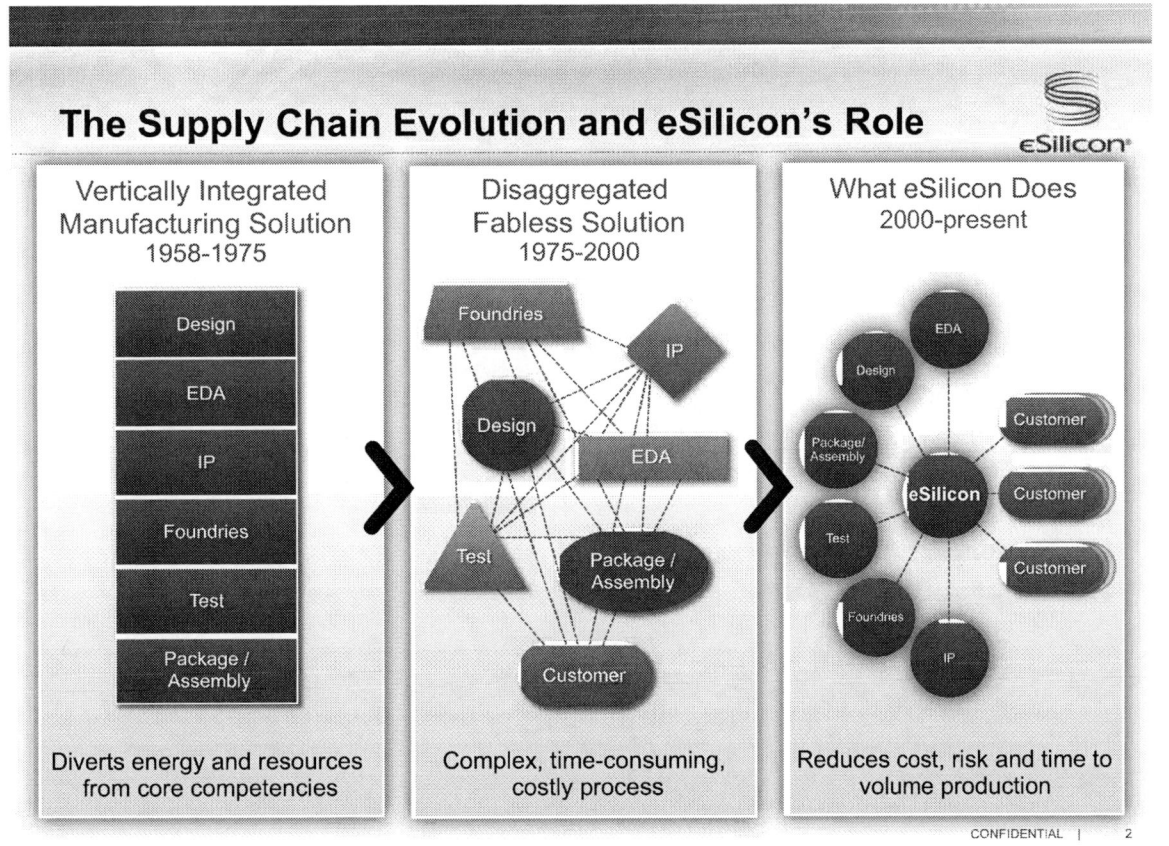

Roadmap and Supply Chain Complexity

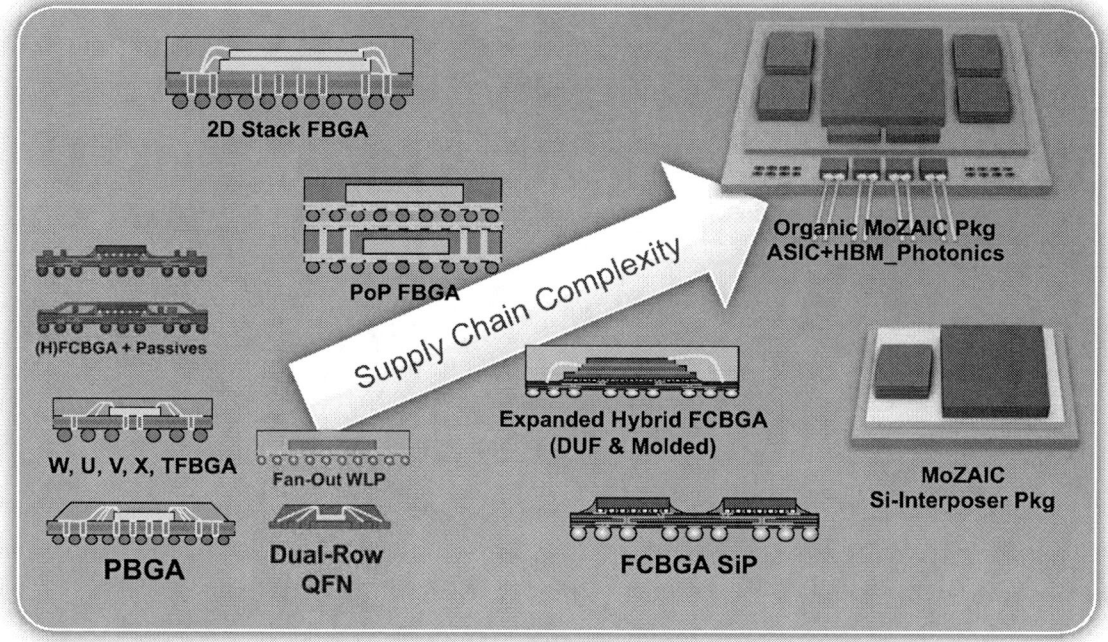

CONFIDENTIAL | 3

Growing Complexity of the Supply Chain

More acute with more integration

- Complex SiP and especially 2.5D and 3DIC solutions require
  - Higher level of co-design of package and various die
  - Sources of die are from different suppliers
  - A role change for IP providers
  - Higher burden of support from supply chain partners such as foundries and OSATs
  - More levels and participants in supply chain
    - Interposer
    - Tiles
    - Photonics
  - More complex yield management and uncertainty of ownership

CONFIDENTIAL | 4

Amkor Technology

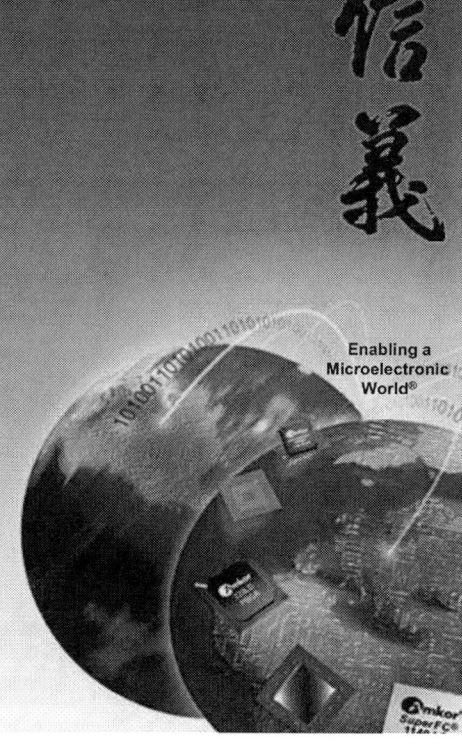

Enabling a Microelectronic World®

R. Huemoeller

SVP, Adv. Product / Platform Develop

September 2013

Amkor Technology

- **Founded 1968**

- **$3B in annual revenues**

- **Largest OSAT package product portfolio**

- **Recognized industry technology leader**

- **Turnkey capabilities**
  - Package and test platform development
  - Volume manufacturing
  - Package design and characterization
  - Finished product logistics

- **Solutions to meet end market demands**

LTM 1Q13 Flip Chip & Advanced Packaging Revenue

($ in millions)

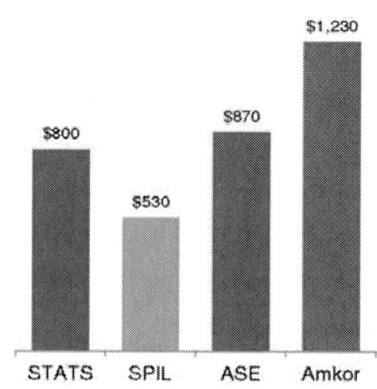

STATS	SPIL	ASE	Amkor
$800	$530	$870	$1,230

Amkor Info for Controlled Release at MEPTECH   2   Sep-13, R.Huemoeller

World Wide Infrastructure and Support

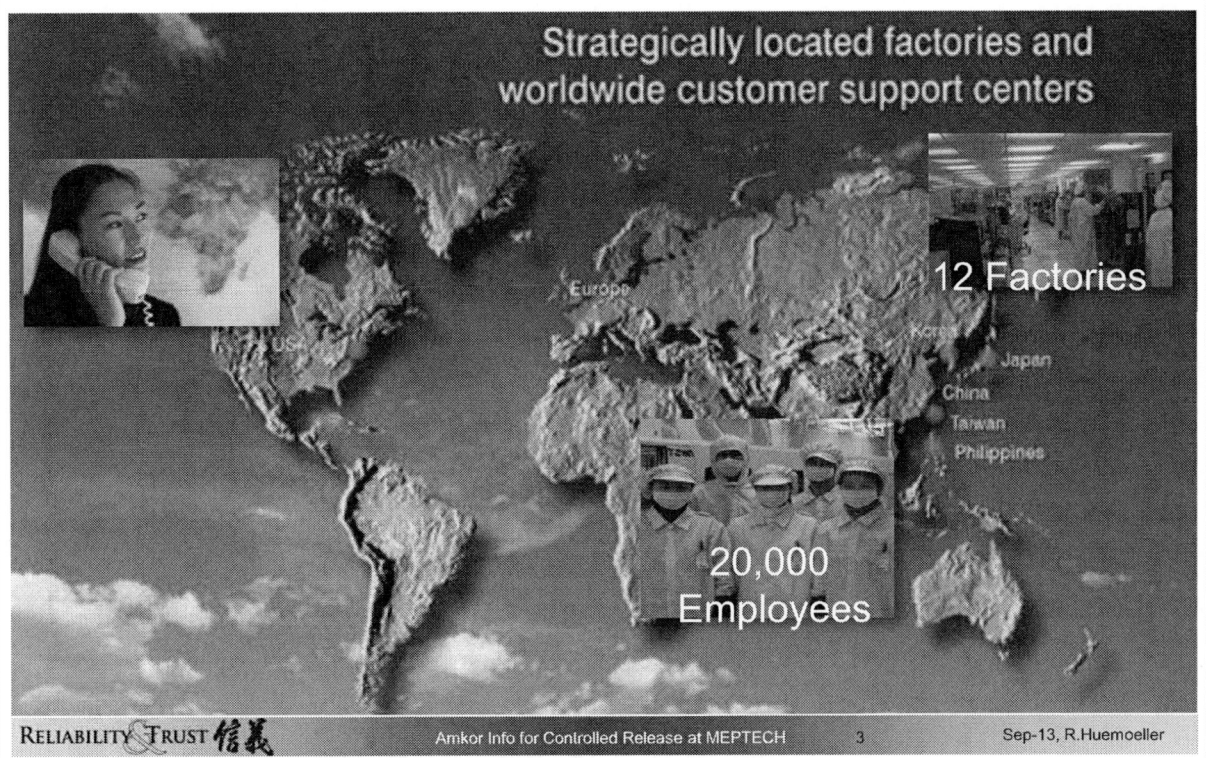

A History of Product & Platform Innovation

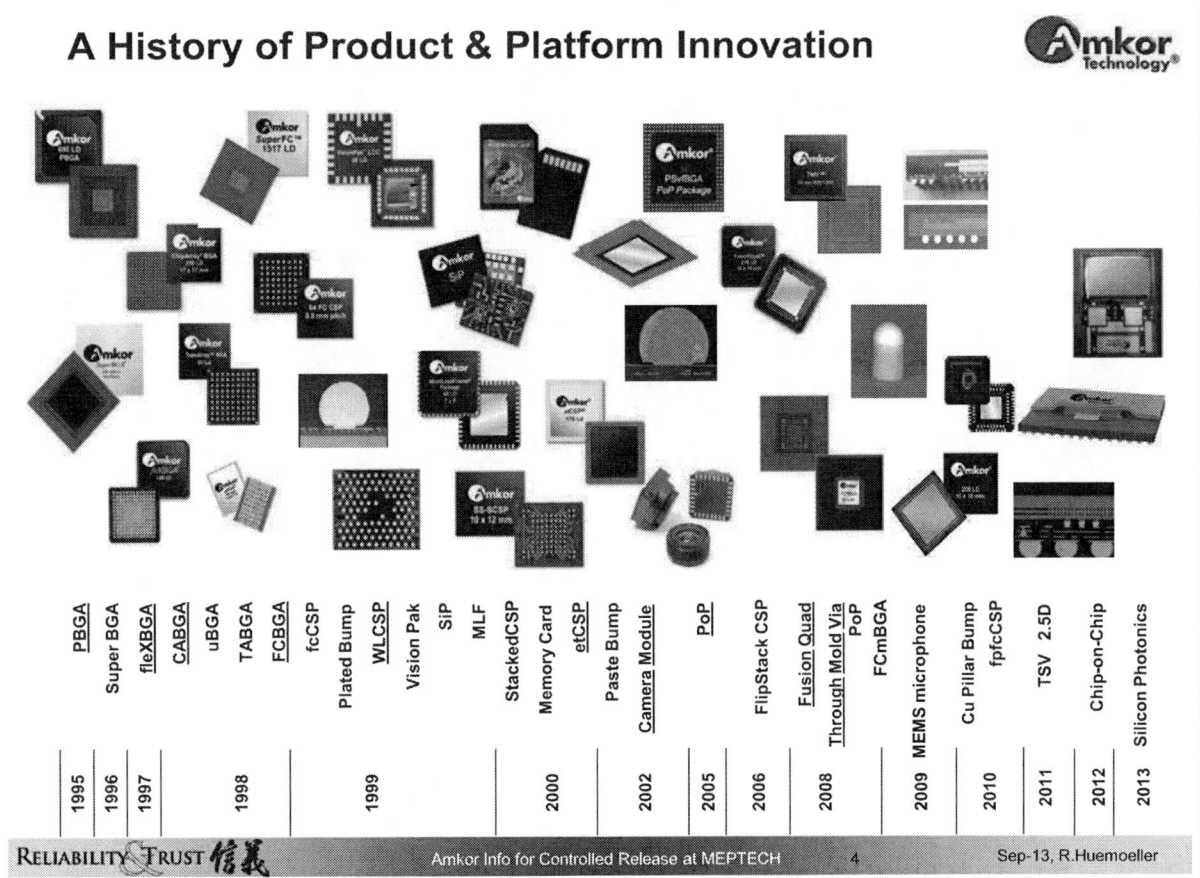

Long Term Advanced Packaging Roadmap

  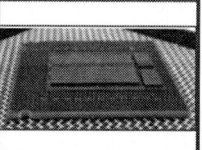  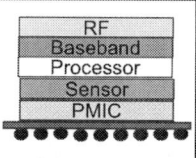

2.5D TSV 20nm Partitioning Cu Pillar C-o-C	Memory Cube Package Stacking C-o-C	2.5D GPU + HBM TSV Embedded Die	2.5D TSV 14nm Partitioning APU SoC Deconstruct	3D TSV Heterogeneous Die Stacking Si Photonics
Market:	**Market:**	**Market:**	**Market:**	**Market:**
FPGA	Mobile Network	Processors Network Tablet	Processors Network Mobile, Tablet	Processors Network Mobile, Tablet
2013	2014	2015	2016	2017

Time

RELIABILITY & TRUST 信義   Amkor Info for Controlled Release at MEPTECH   5   Sep-13, R.Huemoeller

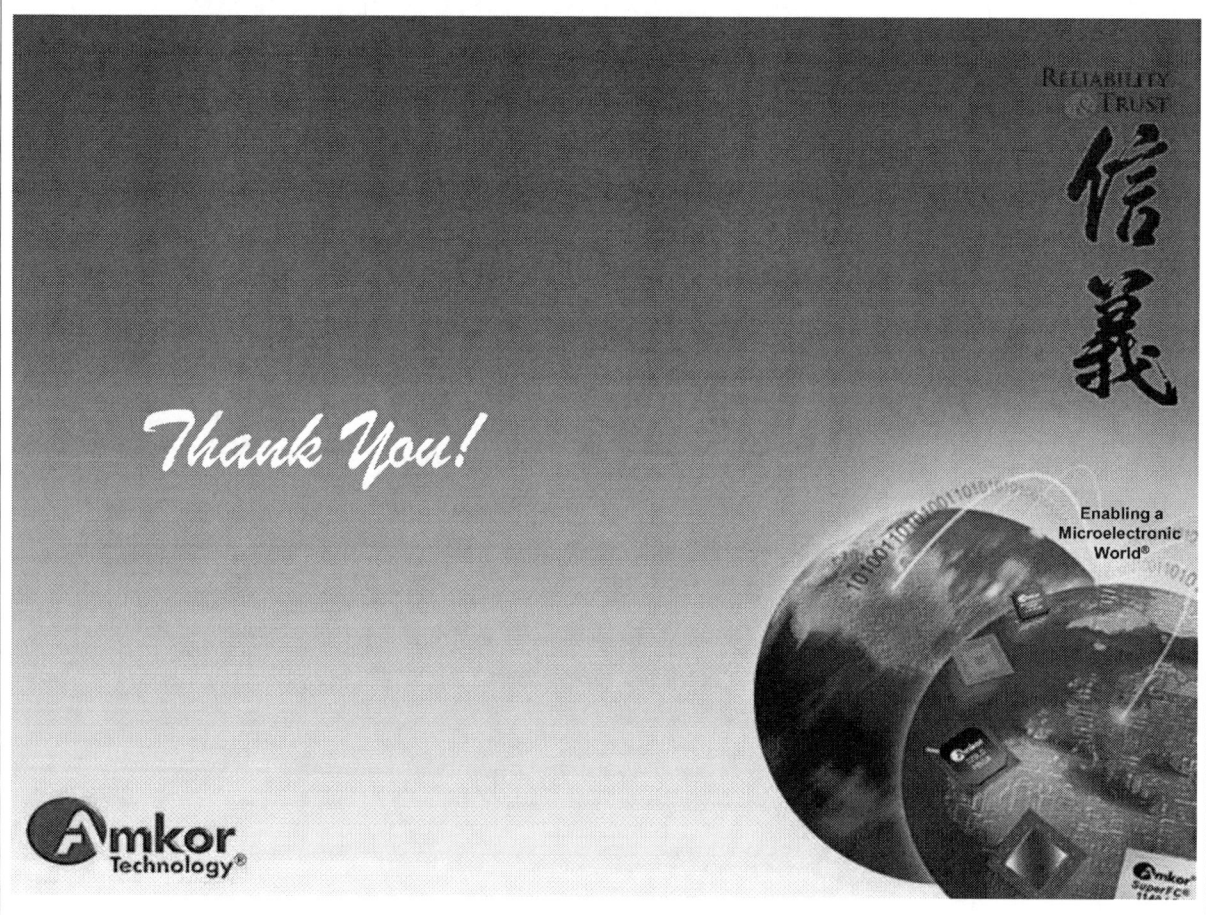

Thank You!

Enabling a Microelectronic World®

Tablets

Gaming

Microsoft

Microsoft Hardware Teams:
30 years of Innovation

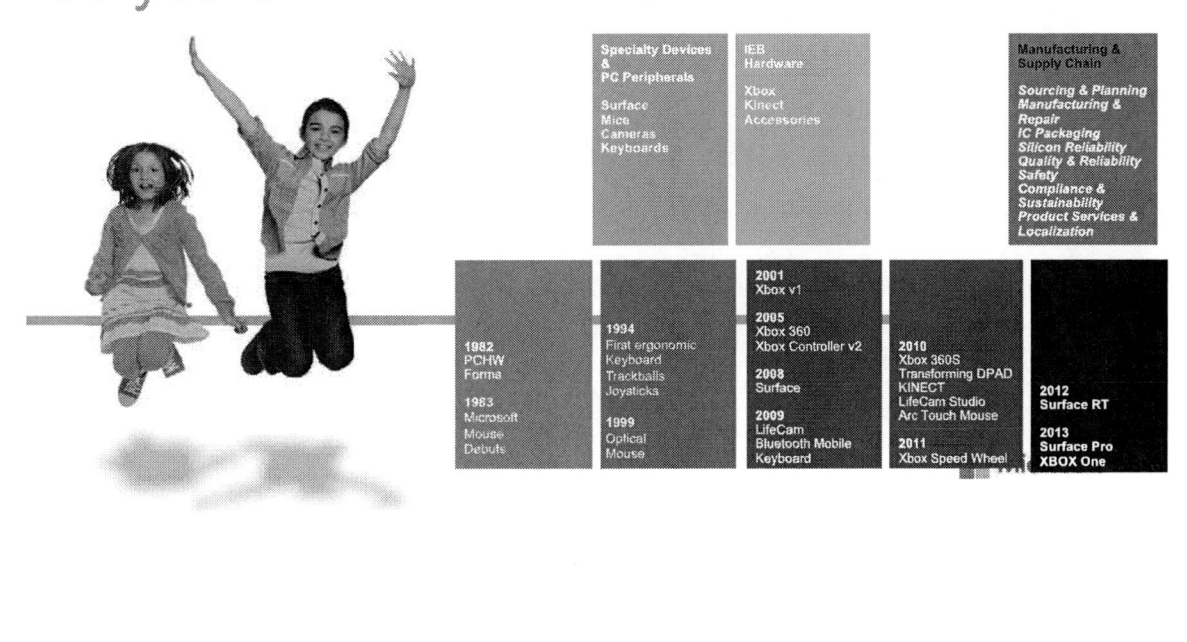

Package Trends: Green Products

- **Hardware Platforms at MSFT have two distinct trajectories**:

1. High performance Processors for Console (XBOX)
   - Long product life cycle – 3 to 5 years
   - Severe thermal management issues. Power > 100W
   - High performance thermal interface materials
   - Die shrinks (process node 28nm >20nm>14nm) resulting in increasing power density.
   - Lower cost processes

2. **Small Form factor driven for handheld devices (Kinnect & Tablets)**
   - Short product development cycle (not field life) of less than one year.
   - Fine pitch devices BGA & lead-frame based devices.
   - Processors with memory integration using Package on Package (POP)
   - Increased use of sensors including TSV packages.
   - Lowest cost implementation

■■ Microsoft

Packaging Trends: Long Product life hardware

Processors   Substrate: 3/2/3 ⟹ Substrate: 2/2/2

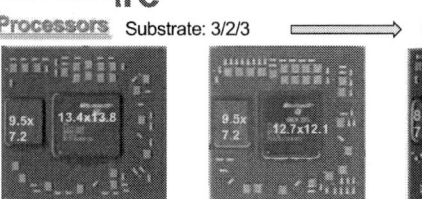

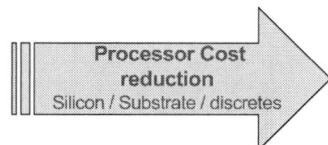

Processor Cost reduction
Silicon / Substrate / discretes

Cameras

NO.	Part Name
1	Protect film
2	Lens
3	Holder
4	Sensor
5	UV glue
6	Rigid-flex

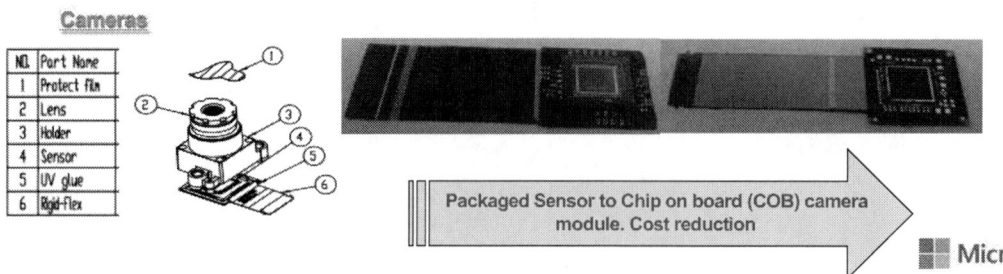

Packaged Sensor to Chip on board (COB) camera module. Cost reduction

■■ Microsoft

Packaging Trends: Short Product life hardware

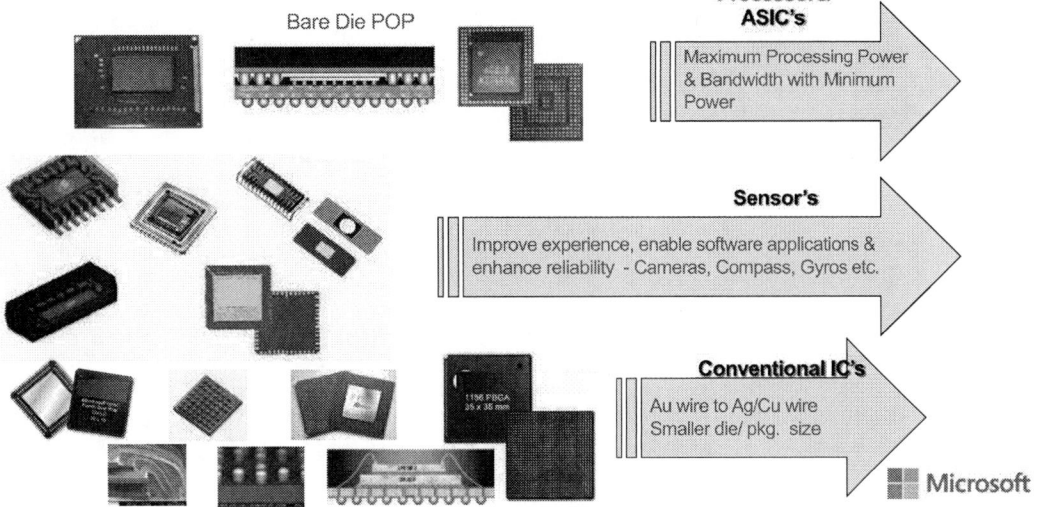

Bare Die POP

Processors/ ASIC's
Maximum Processing Power & Bandwidth with Minimum Power

Sensor's
Improve experience, enable software applications & enhance reliability - Cameras, Compass, Gyros etc.

Conventional IC's
Au wire to Ag/Cu wire
Smaller die/ pkg. size

Microsoft

Summary - Packaging Trends

- **Always Green**

- **Performance**
  Flip Chip development for 20nm & 14nm Flip chip assembly/ Copper pillar
  Aggressive thermal/mechanical management

- **Form Factor**
  Stacked Die and Package stacking (PoP)
  Increased MEMS and sensor packaging

- **Cost**
  Reduction in Gold use
  Aggressive cost reduction roadmap

Microsoft

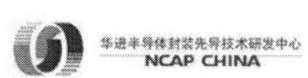

3D Microelectronics Packaging:
Technology Development & Commercialization

Dr. Dongkai Shangguan

CEO – National Center for Advanced Packaging
(NCAP China)

MEPTEC Semiconductor Roadmaps Conference

September 2013

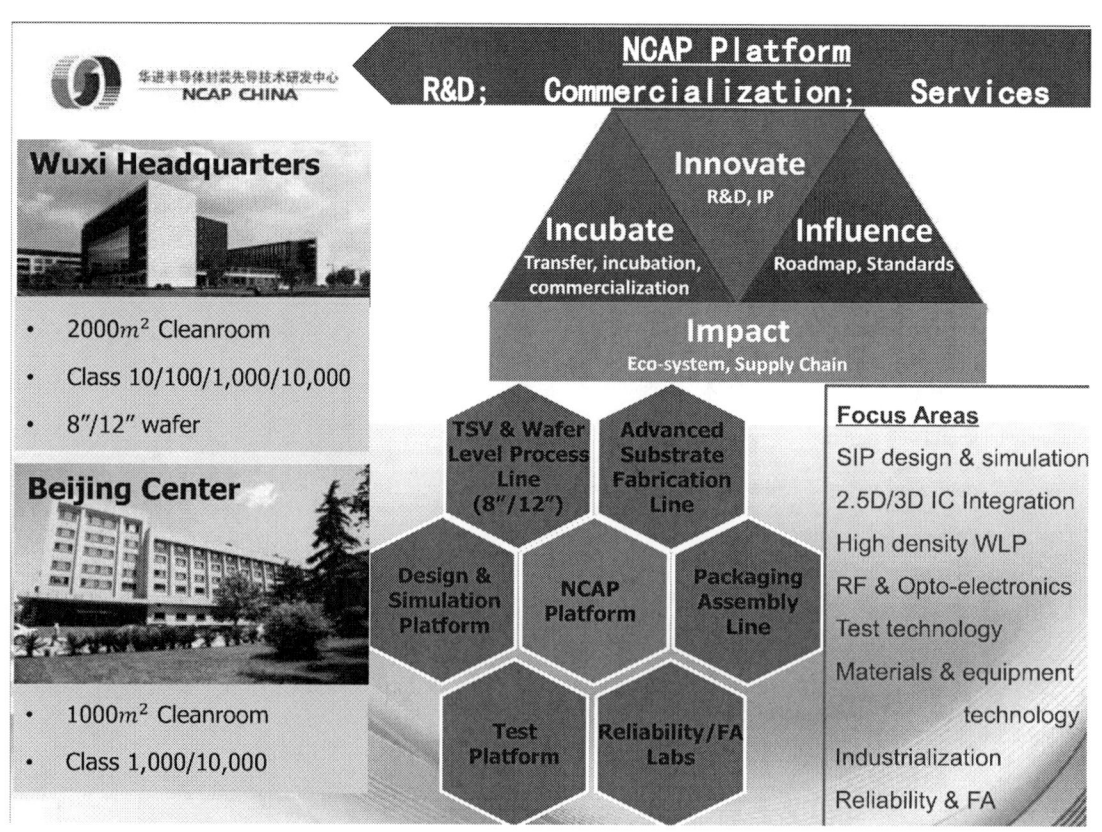

3D SiP: Multi-tiered & Multi-faceted

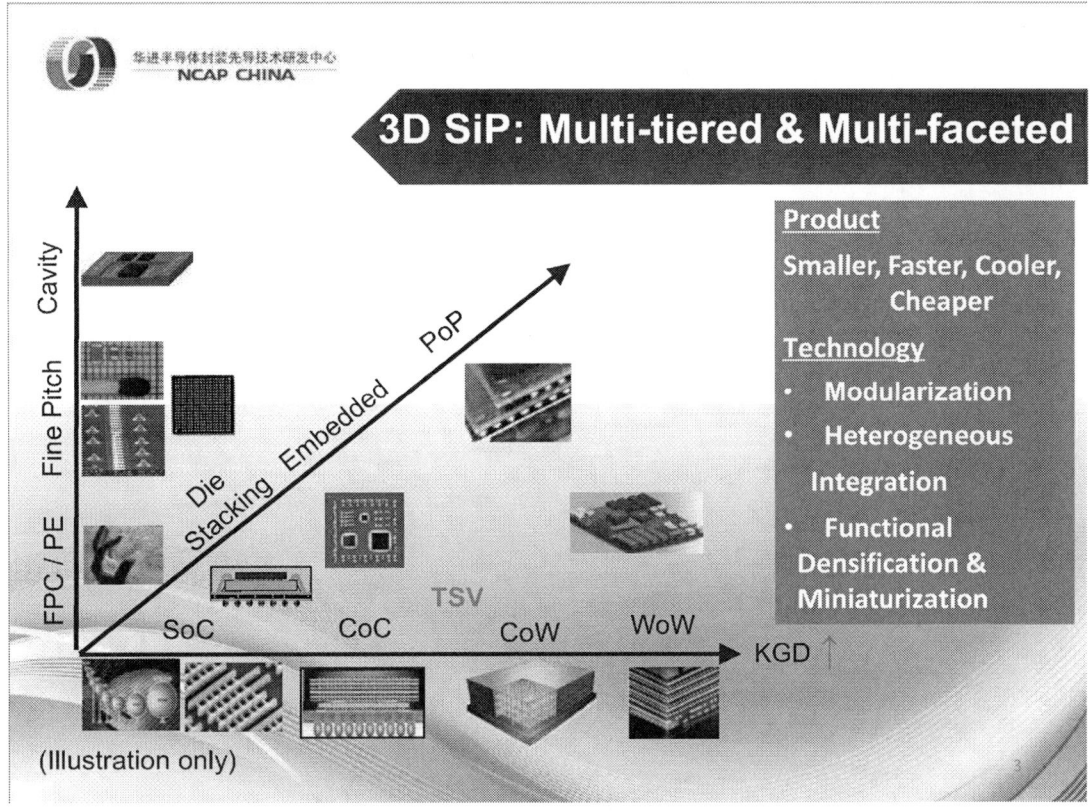

Product
Smaller, Faster, Cooler, Cheaper

Technology
- Modularization
- Heterogeneous Integration
- Functional Densification & Miniaturization

(Illustration only)

Industry Challenges (2.5D/3D)

- Design guidelines & EDA tools
  - Cross domain
- Architecture & partition
- Thermal management
- Mechanical Reliability
- Total optimization

from partitioned solutions to integrated platform solutions

Supply Chain
- IDM / Fabless
- EDA
- Foundries
- OSAT
- Materials & equipment
- R&D
- Standards Bodies

- Cost (system cost, TCoO)
- KGD
- Multiple IC vendors
- Multiple IC technologies
- Process flow & partition
- Inter-operability
- Infrastructure
- Standards & specifications
- Convergence, Integration, and Business Model

- Wafer thinning & handling(<100um)
- TSV forming & filling
- RDL / Microbumps
- Bonding / debonding
- C2C, C2W, W2W
- Underfill

- Equipment & tooling
  Alignment accuracy, …
- Warpage
- Throughput / Yield
- Metrology / Inspection methodology
- Test methodology & equipment (BIST…)

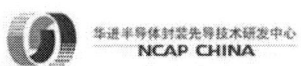

Total Solution
Design
Materials + Equipment + Process
Performance + CoO

The Cost Factor
➢ Currently, cost is still a major barrier – especially for high routing density applications
➢ Current applications are primarily driven by high value/performance products, with limited volume
➢ The main cost factor is equipment, followed by materials
  • Industry-wide collaboration is needed to lower the total cost
➢ Interposer materials: silicon, glass, organic, ceramic

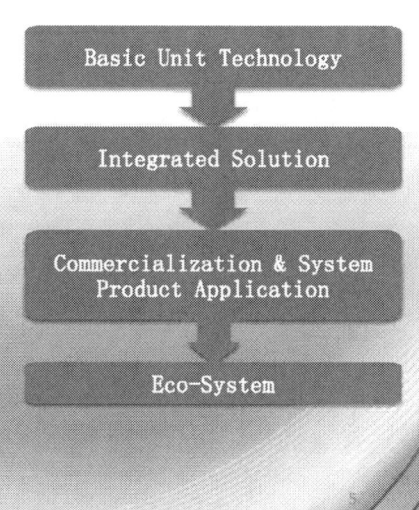

Manufacturing Drivers

Linda Matthew
TechSearch International

FC 2010-03

© 2010 TechSearch International, Inc.

Panelists

- **Moderator: Jeff Demmin, STATS ChipPAC**

- **Richard Crisp, Invensas**
- **Javier DeLaCruz, eSilicon**
- **Ron Huemoller, Amkor**
- **Raj Master, Microsoft**
- **Dongkai Shangguan, NCAP China**

FC 2010-03

© 2010 TechSearch International, Inc.

Summary

- **Highlight issues, recap challenges and solutions**

FC 2010-03 © 2010 TechSearch International, Inc.

Summary

- **TSV and multi-die packaging**
  - **Functional integration and/or SoC deconstruction**
  - **Form-factor**
  - **Power**
- **Photonics**
  - **Bandwidth, latency**
  - **Cisco high performance packaging option**
  - **Amkor roadmap: silicon photonics in 2013**
  - **eSilicon: photonics transceiver on module**
- **Reliability**
  - **500 cycles for mobile phones, going to 400 or 300**

FC 2010-03 © 2010 TechSearch International, Inc.

TSV - Challenges

- **When will TSV arrive? – continues to be delayed – Tom Gregorich**
- **KGD**
    - **Essential element to TSV deployment**
    - **"In my view, we haven't made much progress" – Raj Master**
    - **Redundancy/repair mature in memory, first gen in logic**
    - **Probe a sub-set of bumps**
    - **"Massive effort, very complex, existing methodologies are broken" – Javier DeLaCruz**

FC 2010-03   © 2010 TechSearch International, Inc.

TSV - Challenges

- **Supply chain/business model**
    - **"Who is going to be the integrator?" – Dieter Bergman**
    - **Yield management is bigger issue than how to build it**
    - **Foundry does packaging ("How well that will work, how long that will last, remains to be seen" – Dongkai Shangguan)**
    - **OSAT does foundry**
    - **eSilicon: integrators tie supply chain together for customers. Will have to take a larger role because ecosystem is becoming too complex**
    - **Supply chain for interposers will evolve and mature, just as supply chain for organic substrates did. – Dongkai Shangguan**

FC 2010-03   © 2010 TechSearch International, Inc.

TSV - Challenges

- 100% yield is not a reasonable target on large 2.5D or 3D packages. Instead well-understood yields are critical. It is not unrealistic to have 60% yield on a very large packaged system.
  – Javier DeLaCruz

- Interposer suppliers and OSATs should not be looked to for tying together the ecosystem. They are not interested in inventory, sourcing of active circuitry and tested yield management.
  – Javier DeLaCruz

- The big gorilla in the room is yield analysis and how it can be done with bump pitches that are so fine. While it can be done, it has to be done much differently and with a higher test escape rate compared to what can be done on much larger bump pitches.
  – Javier DeLaCruz

Challenges in TSV

- "Don't think memory will be the big play for early implementation of TSV – it will be the de-partitioning of logic at the 14 and 16nm nodes" – Ron Huemoller
  - 2015: High volume 2.5D starting, with availability of HBM
  - Deconstruction of logic for 16/14nm will be primary driver for the value markets as opposed to memory integration
  - 2016/17: True 3D packages with memory on top – delayed by cost

Alternatives to TSV

- "Industry abhors revolutionary change" – Richard Crisp
- Alternatives for modularity:
  - Extensions of PoP
  - Multichip WLFO
  - F2F (CoC)

FC 2010-03   © 2010 TechSearch International, Inc.

Solutions in TSV

- Availability of HBM
- NCAP China: development of non-wafer fab equipment optimized for packaging
- "We hope to come up with lower cost solutions to enable us to deploy 2.5D solutions" – Dongkai Shangguan
- eSilicon: integrators tie supply chain together for customers.
- Supply chain for interposers will evolve and mature, just as supply chain for organic substrates did. – Dongkai Shangguan

FC 2010-03   © 2010 TechSearch International, Inc.

The Collaboration Engine:
Enabling Innovation in Microelectronics

Karen Savala
President, SEMI Americas

Outline

- About SEMI
- Semiconductors: A History of Collaboration
- Collaboration in other Microelectronics Industries
- 3D IC Crossroads

SEMI: The Global Association

- Global Association
    - ~1800 member companies
- Events
    - SEMICON, SOLARCON
- Standards
- Advocacy
- Market Research
- EHS
- Semiconductor, LED, FPD, MEMS, PV, plastic electronics, emerging markets

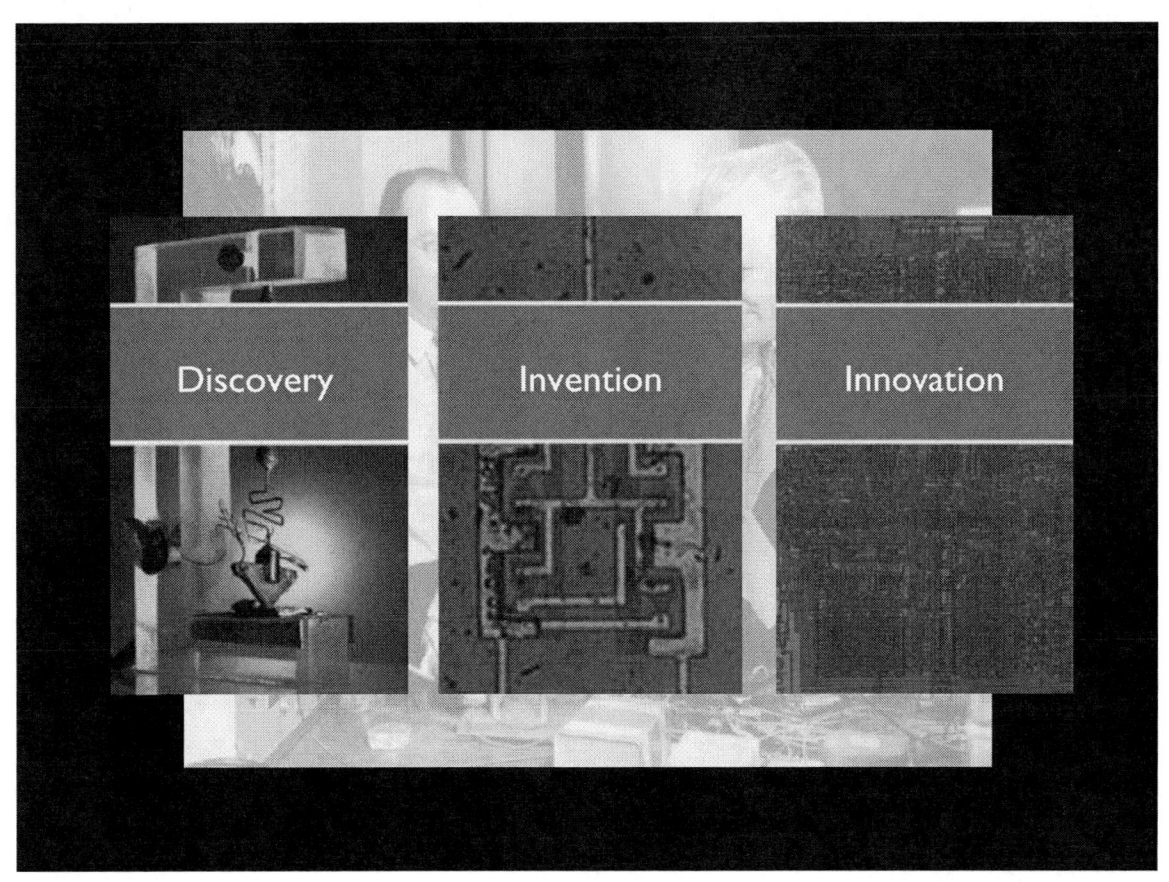

Semiconductors:
A History of
Collaboration

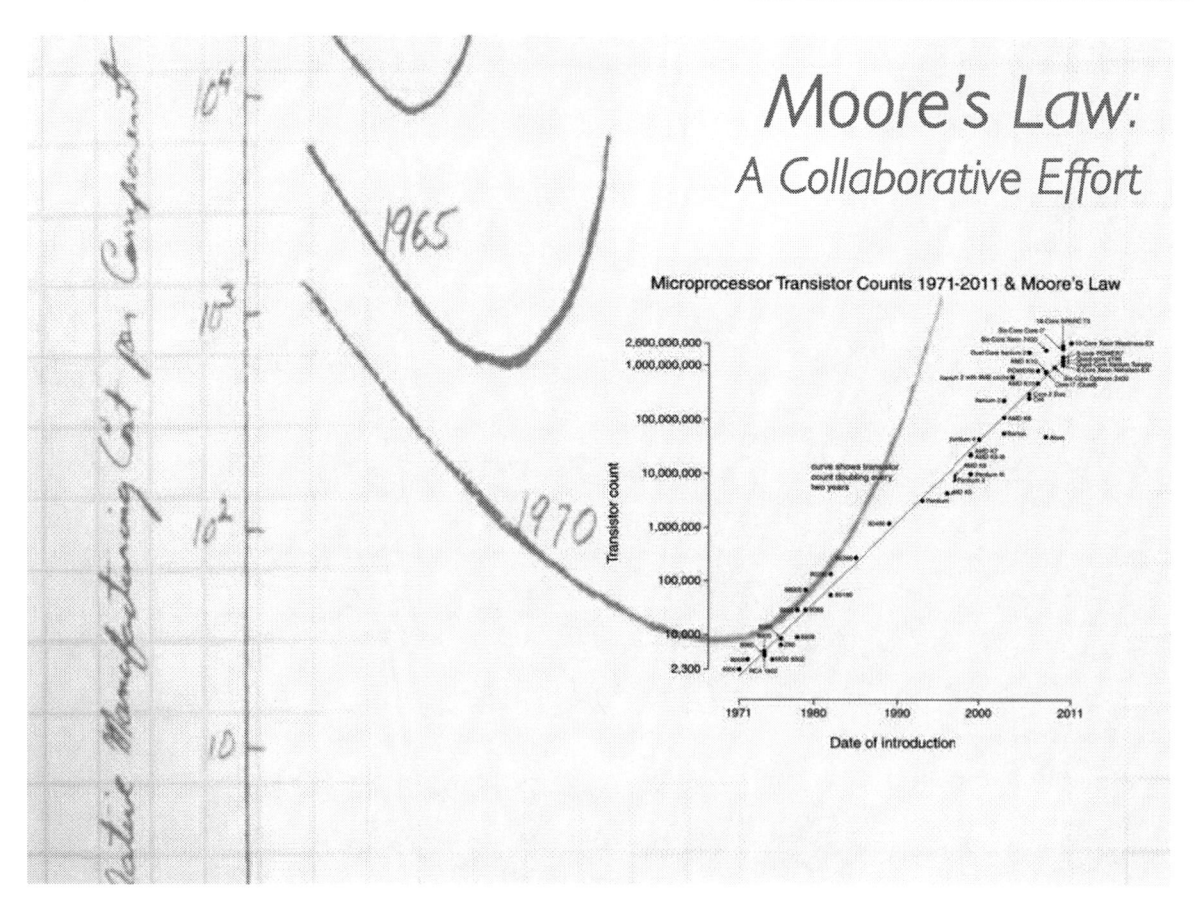

R&D Collaboration: A New Era in Industry Collaboration

Manufacturing Standards: Reducing Cost, Enabling Innovation

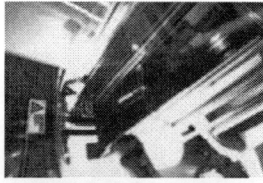
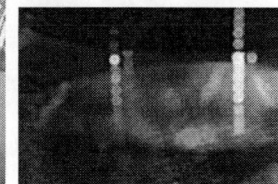

SEMI International Standards

Collaboration:
Critical to Scale in Emerging Industries

Solar PV: Standards to Support a Global Trillion Dollar Industry

SEMI PV Standards Organization

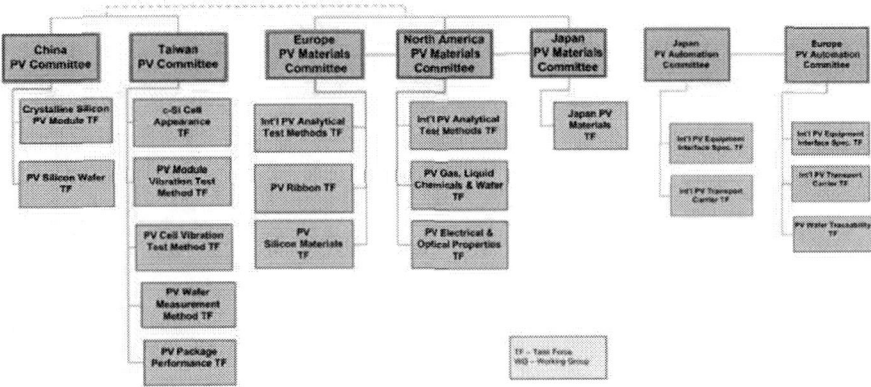

International PV Technology Roadmap (ITRPV)

- ITRPV has become a major contributor to identifying and focusing the industry on key cost reduction opportunities in c-Si PV
- Now in its 3rd Edition
- Participation from leading suppliers and manufacturers in US, Asia and Europe
- Cost reduction targets identified in 18 areas

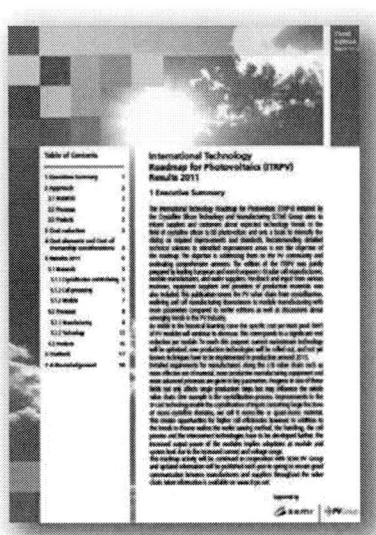

LED: Powering the Solid State Lighting Revolution

- Critical technology for reduction of energy consumption
- Standards: leading the way to reducing costs, accelerating technology development and adoption

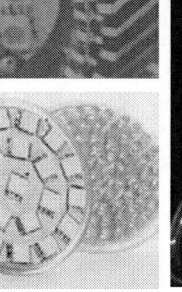
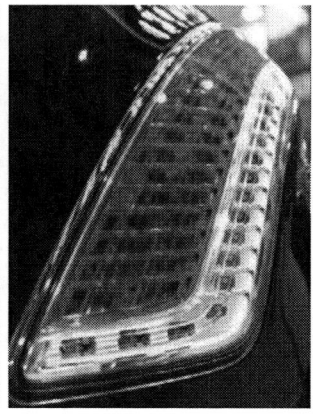

LED Standards: Reducing Costs to Enable Solid State Lighting

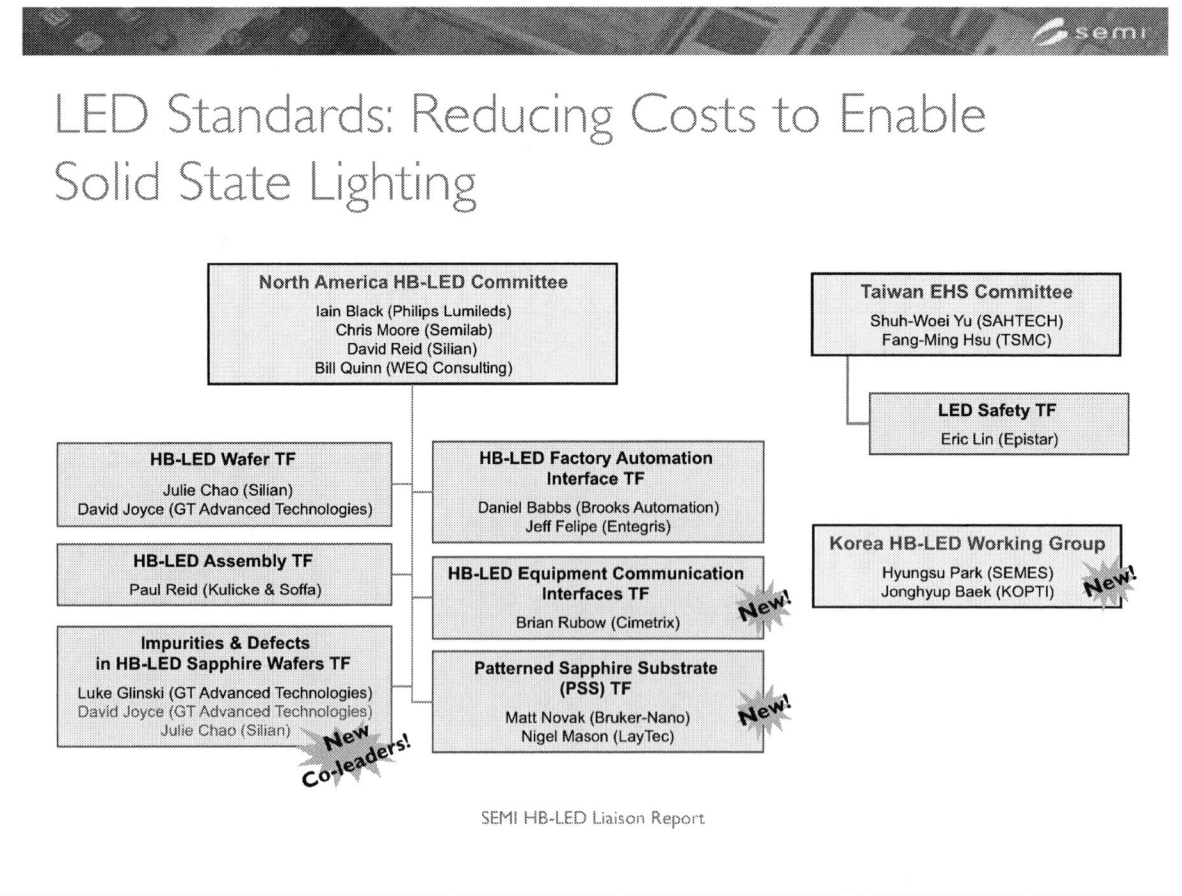

North America HB-LED Committee
Iain Black (Philips Lumileds)
Chris Moore (Semilab)
David Reid (Silian)
Bill Quinn (WEQ Consulting)

Taiwan EHS Committee
Shuh-Woei Yu (SAHTECH)
Fang-Ming Hsu (TSMC)

LED Safety TF
Eric Lin (Epistar)

HB-LED Wafer TF
Julie Chao (Silian)
David Joyce (GT Advanced Technologies)

HB-LED Factory Automation Interface TF
Daniel Babbs (Brooks Automation)
Jeff Felipe (Entegris)

HB-LED Assembly TF
Paul Reid (Kulicke & Soffa)

HB-LED Equipment Communication Interfaces TF
Brian Rubow (Cimetrix)

New!

Korea HB-LED Working Group
Hyungsu Park (SEMES)
Jonghyup Baek (KOPTI)

New!

Impurities & Defects in HB-LED Sapphire Wafers TF
Luke Glinski (GT Advanced Technologies)
David Joyce (GT Advanced Technologies)
Julie Chao (Silian)

New Co-leaders!

Patterned Sapphire Substrate (PSS) TF
Matt Novak (Bruker-Nano)
Nigel Mason (LayTec)

New!

SEMI HB-LED Liaison Report

New Industries Based on New Materials

**Manufacturing Standards:
Enabling Innovation**

450mm Standards:
Essential for Innovation

- SEMI: advocating open process, facilitating readiness through standards
- Transition costs: high
    - US$8-$30 billion
- ✓ Collaboration essential

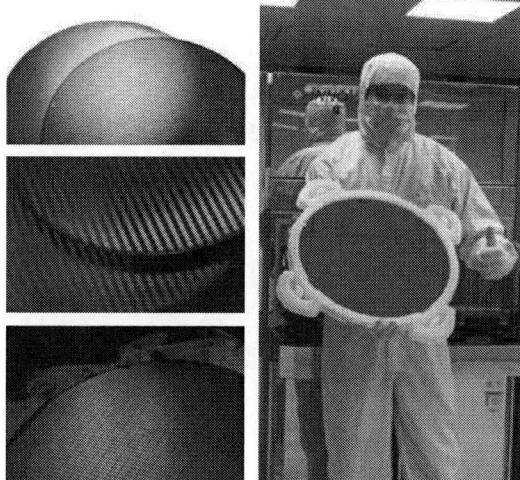

Stacked ICs Have Been Around Awhile...

1979: Mostek--MK2716 EPROM riding piggyback on top of the MK38P70 microcontroller.

Source: EDA360 Insider, Nov. 9, 2011

1985: Texas Instruments:--Two, 64Kbit, plastic-packaged DRAMs in DIPs have been stacked and soldered to produce a 128Kbit device.

SEMI 3DS-IC Standard Organization Chart

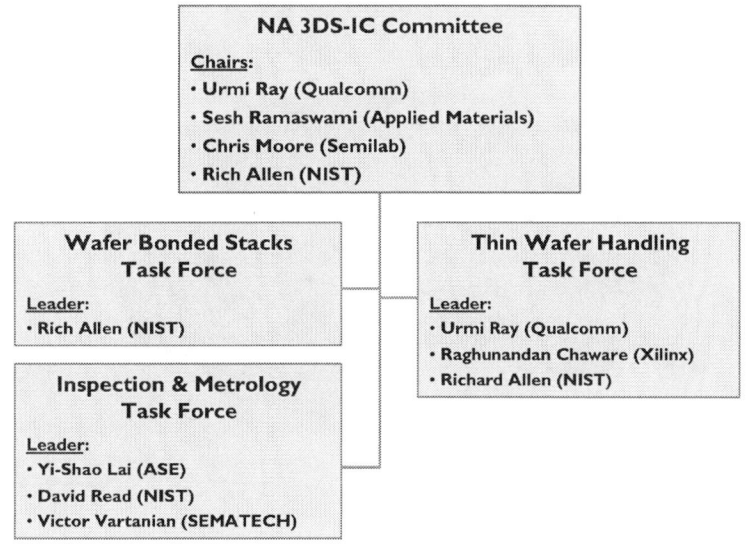

NA 3DS-IC Committee

<u>Chairs:</u>
- **Urmi Ray (Qualcomm)**
- **Sesh Ramaswami (Applied Materials)**
- **Chris Moore (Semilab)**
- **Rich Allen (NIST)**

Wafer Bonded Stacks Task Force

<u>Leader:</u>
- **Rich Allen (NIST)**

Thin Wafer Handling Task Force

<u>Leader:</u>
- **Urmi Ray (Qualcomm)**
- **Raghunandan Chaware (Xilinx)**
- **Richard Allen (NIST)**

Inspection & Metrology Task Force

<u>Leader:</u>
- **Yi-Shao Lai (ASE)**
- **David Read (NIST)**
- **Victor Vartanian (SEMATECH)**

SEMI 3DS-IC Standards Activities - July 2013

3DS-IC: Technical Challenges

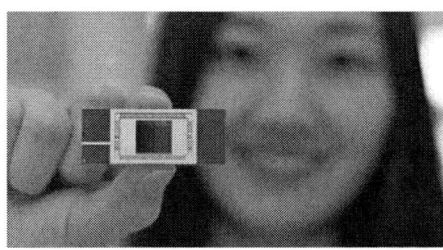

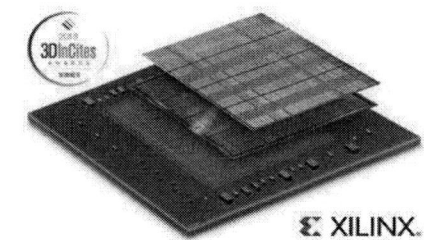

Samsung Ships First Vertical 3D Memory August 2013, Engadget

Xilinx Wins 2013 3D InCites Award for World's First Heterogeneous All Programmable 3D IC (July 2013, PR Newswire)

3D IC Potential

2017 Unit Shipments

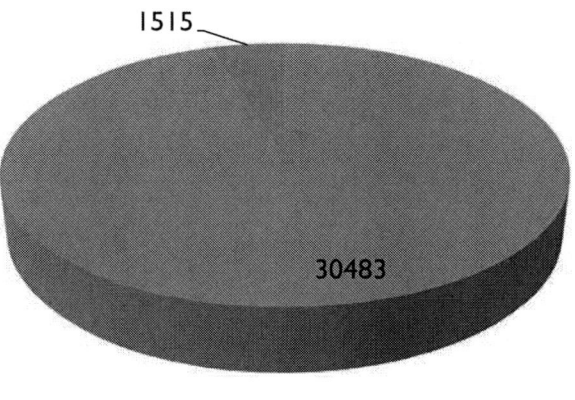

1515

30483

▪ WLP ▪ TSV

In Million Units
Gartner 2013
Excludes CMOS Sensors and MEMS

Competing Business Models

TSMC path

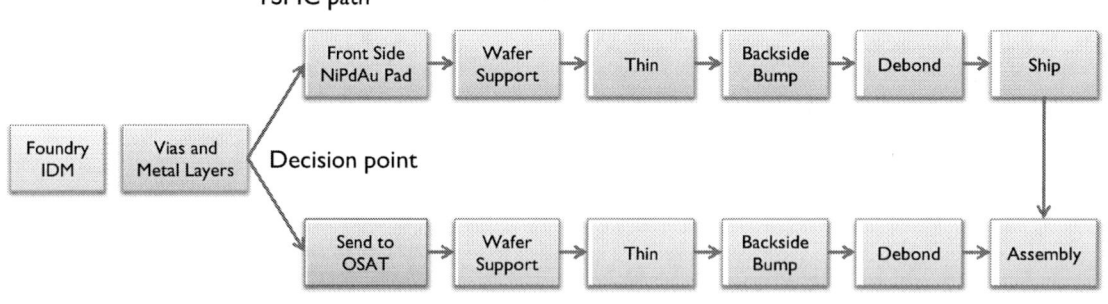

Decision point

UMC and GLOBALFOUNDRIES path

Ref: R. Huemoeller, Amkor Technology

The Next Step

- Meaningful collaboration needed across value chain
- Standards work helps, but new model needed to make 3D IC affordable

Semiconductor • PV • Emerging Markets

Driving the Electronics Revolution

Panel Three: Electrical Performance Requirements to meet Emerging Interconnect Standards

Moderators:

Ivor Barber - Xilinx Inc. and John Xie - Altera Corp.

Panelists:

- Tom Gregorich - Broadcom
- Brad Griffin - Cadence Design Systems
- Anthony Torza - Xilinx Inc.
- Abe Yee - NVIDIA Corporation

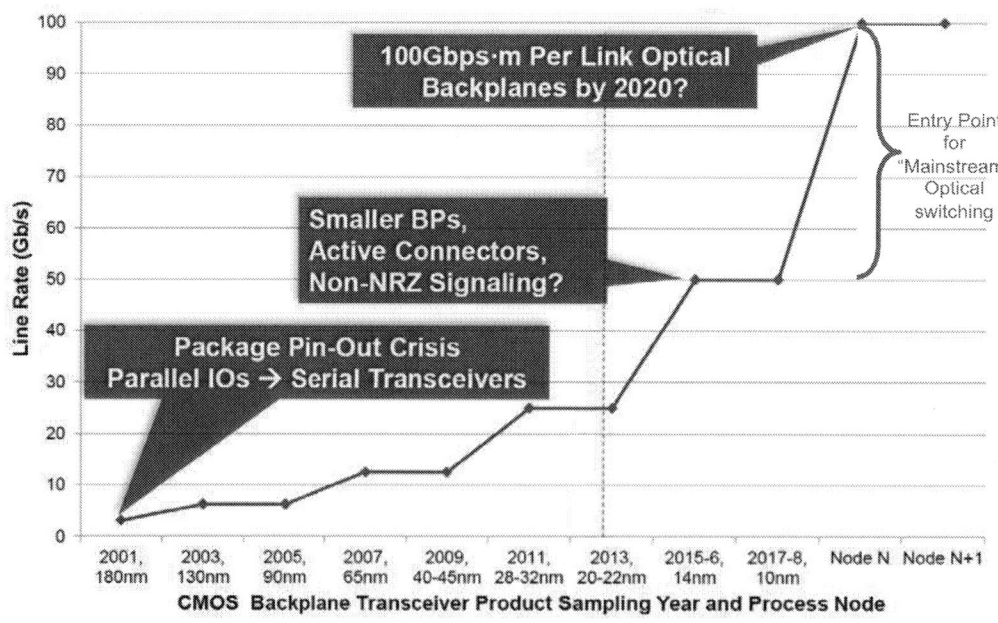

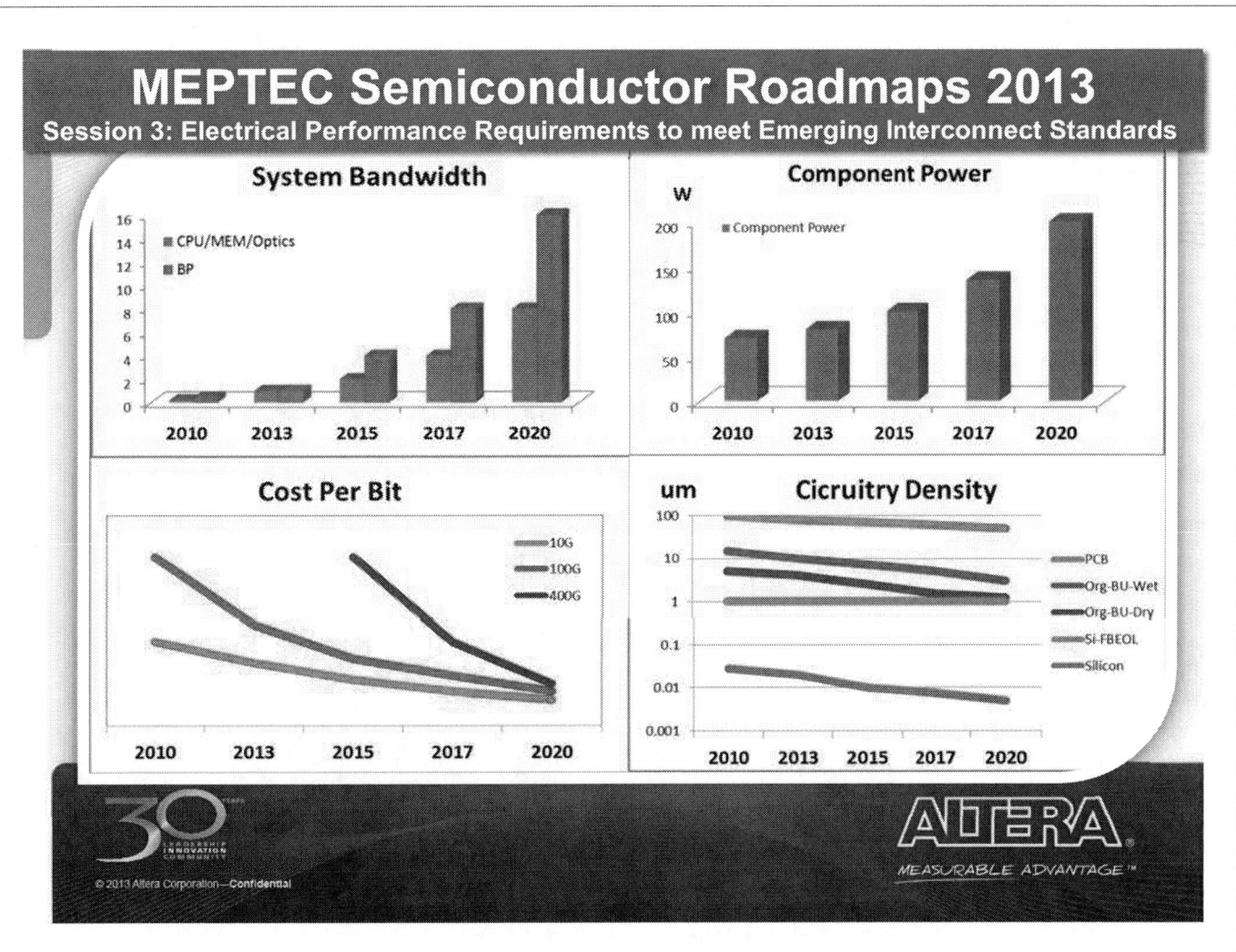

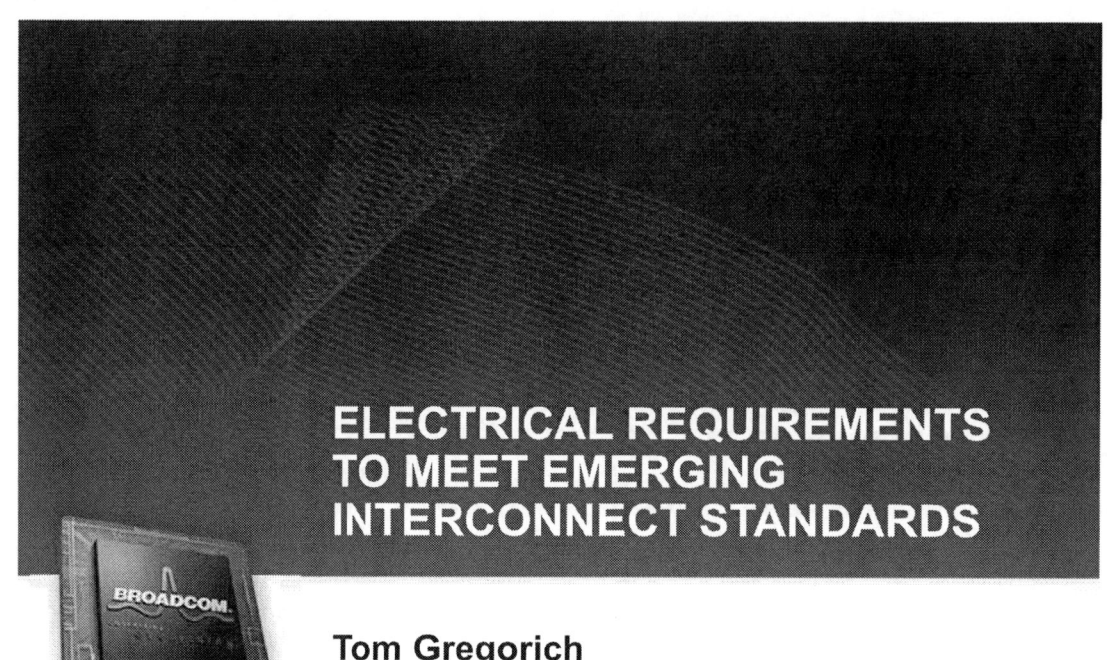

ELECTRICAL REQUIREMENTS TO MEET EMERGING INTERCONNECT STANDARDS

Tom Gregorich
Director, IC Package Engineering
2013 Semiconductor Roadmap Symposium
September 24, 2013

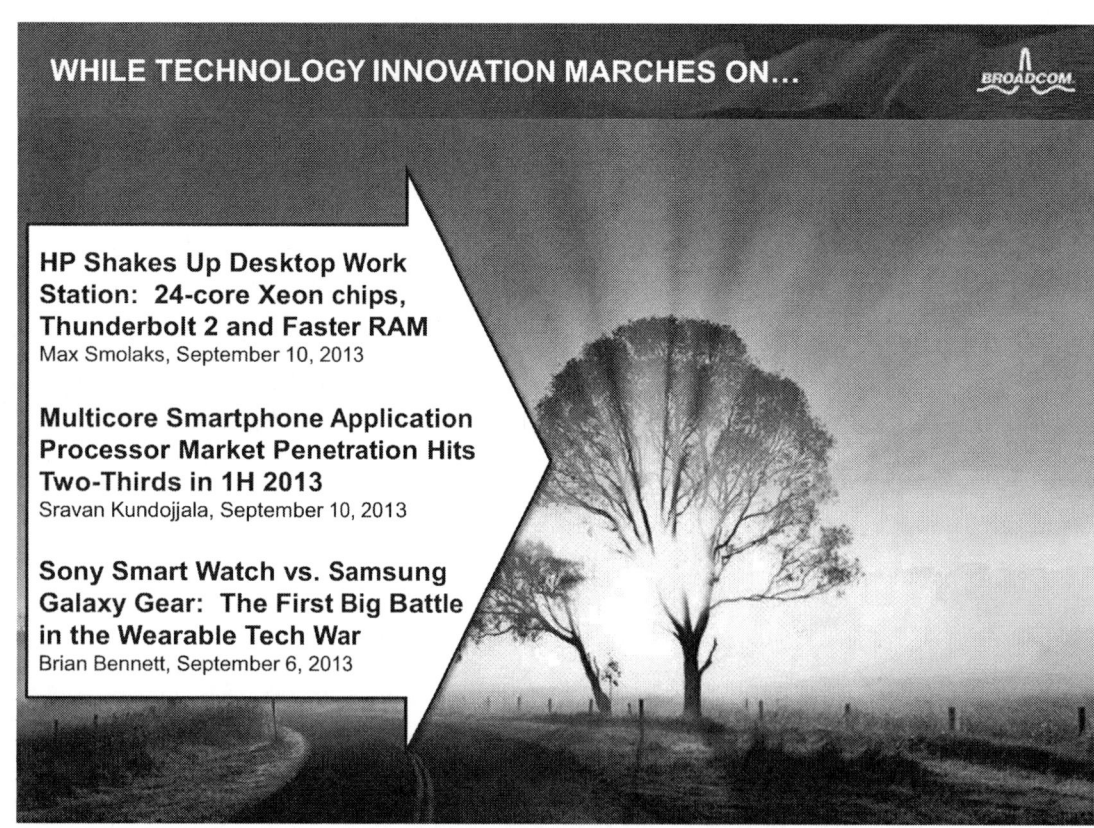

WHILE TECHNOLOGY INNOVATION MARCHES ON...

HP Shakes Up Desktop Work Station: 24-core Xeon chips, Thunderbolt 2 and Faster RAM
Max Smolaks, September 10, 2013

Multicore Smartphone Application Processor Market Penetration Hits Two-Thirds in 1H 2013
Sravan Kundojjala, September 10, 2013

Sony Smart Watch vs. Samsung Galaxy Gear: The First Big Battle in the Wearable Tech War
Brian Bennett, September 6, 2013

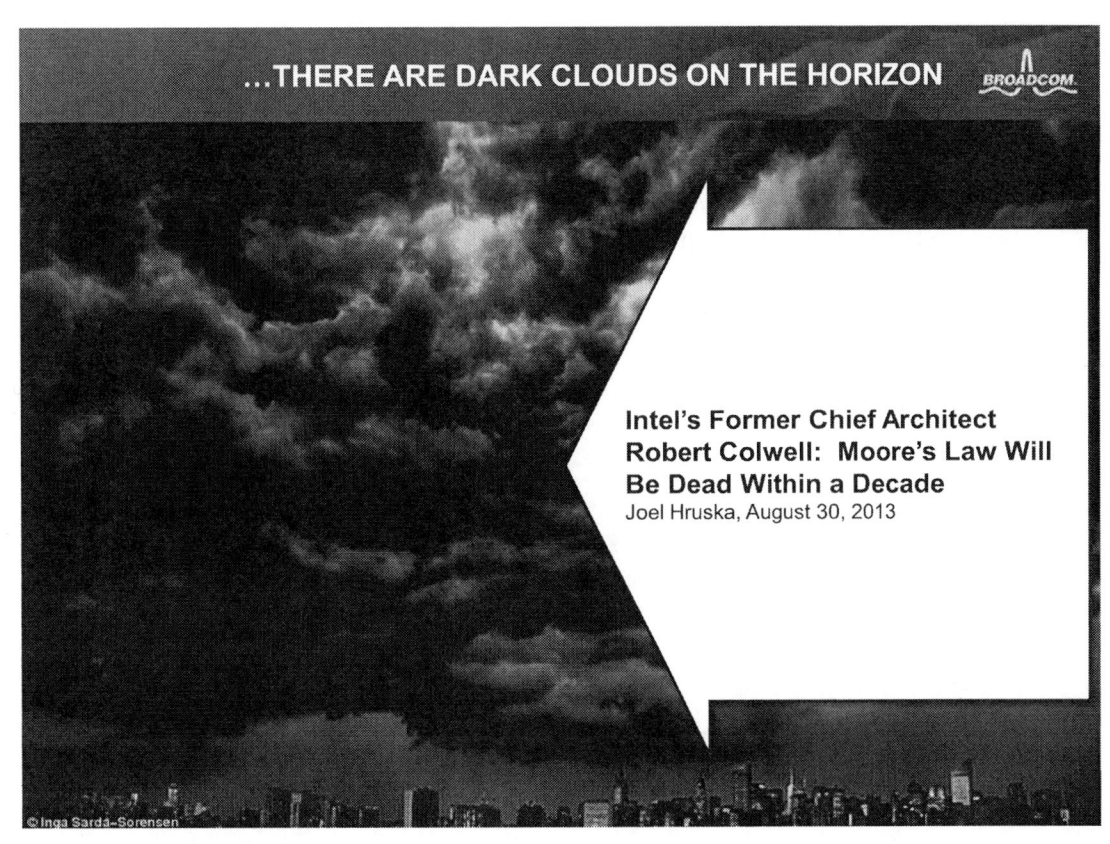

...THERE ARE DARK CLOUDS ON THE HORIZON

Intel's Former Chief Architect Robert Colwell: Moore's Law Will Be Dead Within a Decade
Joel Hruska, August 30, 2013

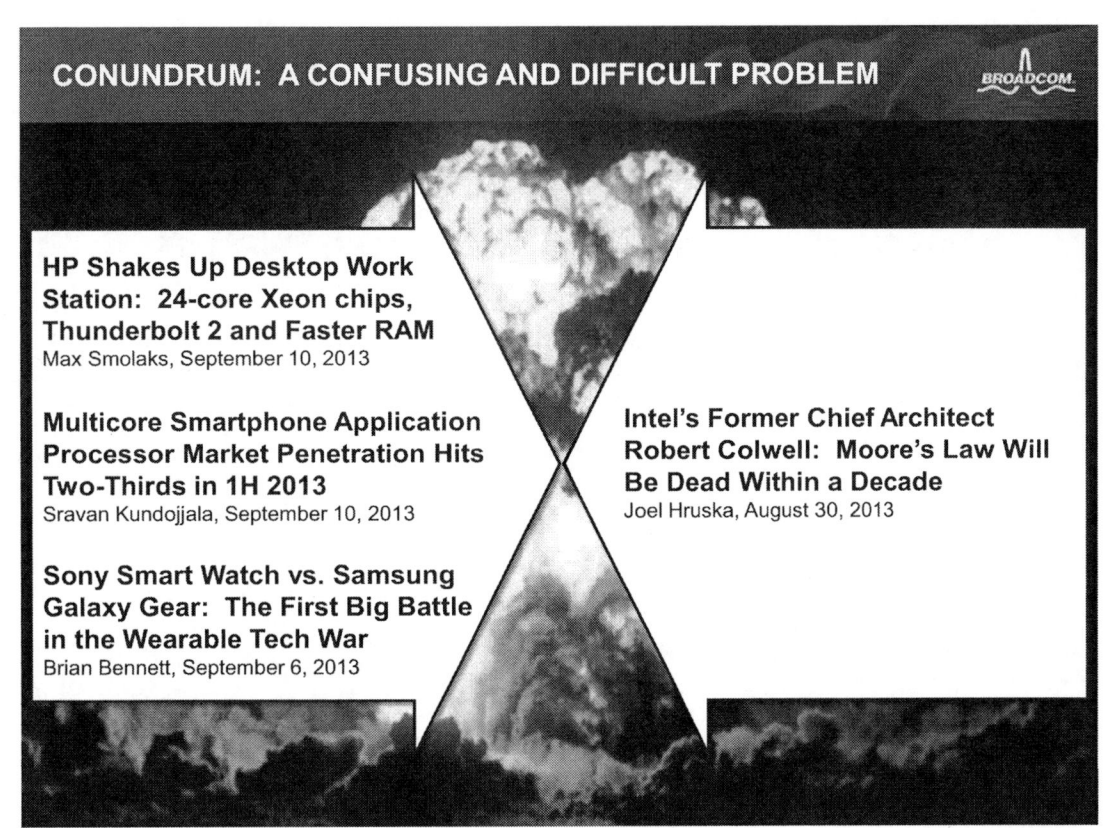

CONUNDRUM: A CONFUSING AND DIFFICULT PROBLEM

HP Shakes Up Desktop Work Station: 24-core Xeon chips, Thunderbolt 2 and Faster RAM
Max Smolaks, September 10, 2013

Multicore Smartphone Application Processor Market Penetration Hits Two-Thirds in 1H 2013
Sravan Kundojjala, September 10, 2013

Sony Smart Watch vs. Samsung Galaxy Gear: The First Big Battle in the Wearable Tech War
Brian Bennett, September 6, 2013

Intel's Former Chief Architect Robert Colwell: Moore's Law Will Be Dead Within a Decade
Joel Hruska, August 30, 2013

KEY QUESTIONS

- What portion of the semiconductor industry's technological success has been a result of Moore's Law?

- Are there alternative technologies which could be used to supplement the benefits currently provided by Moore's Law?

- How much of the technology gap resulting from the slowing of Moore's Law can be filled by these alternative technologies?

- What technological, organizational and systemic changes will be needed to enable use of these alternative technologies?

OPINIONS AND SUGGESTIONS

There are probably no singular replacements for Moore's Law and probably no silver bullets

Successful 3D TSV products must be preceded by successful 3D products using other 3D technologies

Winning strategies will rely on relentless study and optimization of partitioning, architecture and customization to enhance value at the system-level

As software evolves it needs to contribute to system efficiency rather than consume it

Study 2nd tier and mixed-signal suppliers to understand how to survive without depending primarily on Moore's Law

Remember "Jonas' Law" (swim fast, eat everything around you or be eaten!)

IC Package Design Trends
Electrical Assessment – Early and Often

Brad Griffin, Product Marketing Director
MEPTEC – Panel
Santa Clara, CA
September 24, 2013

cādence®

- Design Complexity – Rising
- Time to Market – Decreasing
- Design Team Size - Shrinking

- **IC Package designers face many different requirements**
  - Marketing, Manufacturing, Electrical, Thermal, Stress, Chip design, Board design
- **Design teams need to be efficient**
  - Traditional methodology: design, analyze, fix design, re-analyze, repeat … NOT PRACTICAL
- **Imagine juggling on a tightrope while riding on a unicycle**
  - It might be possible, but very error prone
- **EDA is empowering IC Package designers with technology that electrically assesses the design**
  - Reducing iterations with analysis teams
  - No models or EM expertise required

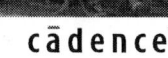

Electrical Performance Assessment (EPA)

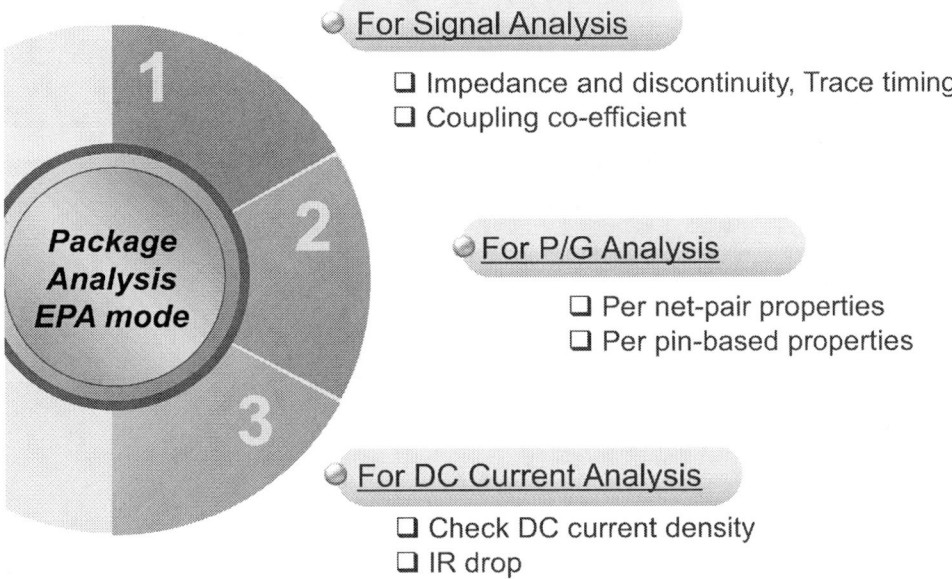

For Signal Analysis
- ❏ Impedance and discontinuity, Trace timing
- ❏ Coupling co-efficient

For P/G Analysis
- ❏ Per net-pair properties
- ❏ Per pin-based properties

For DC Current Analysis
- ❏ Check DC current density
- ❏ IR drop

cādence

Example – Find problematic power and ground pins with a "push of the button"

- 3D displays of R and L
  - Quantify the relative distributions
  - 2D, 3D graphical display - numerical tables can be loaded into spreadsheets

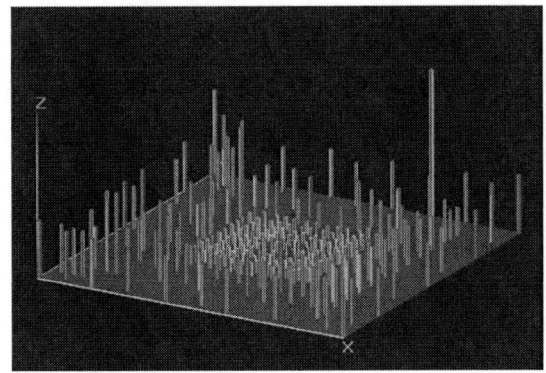

Resistance

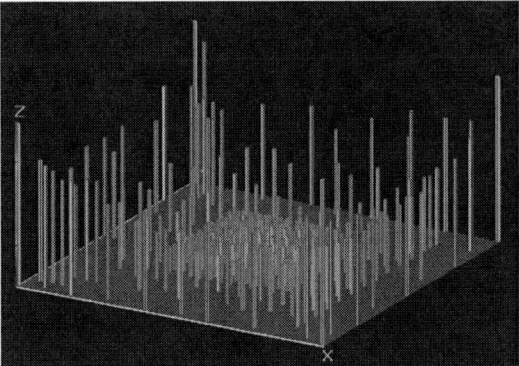

Inductance

cādence

IC Package Electrical Performance Assessment

* It might not solve all the problems for the IC Package designer … but it is a good place to start

cadence®

Next Generation FPGA Packaging Trends

Anthony Torza, Product Planning, Xilinx

© Copyright 2013 Xilinx

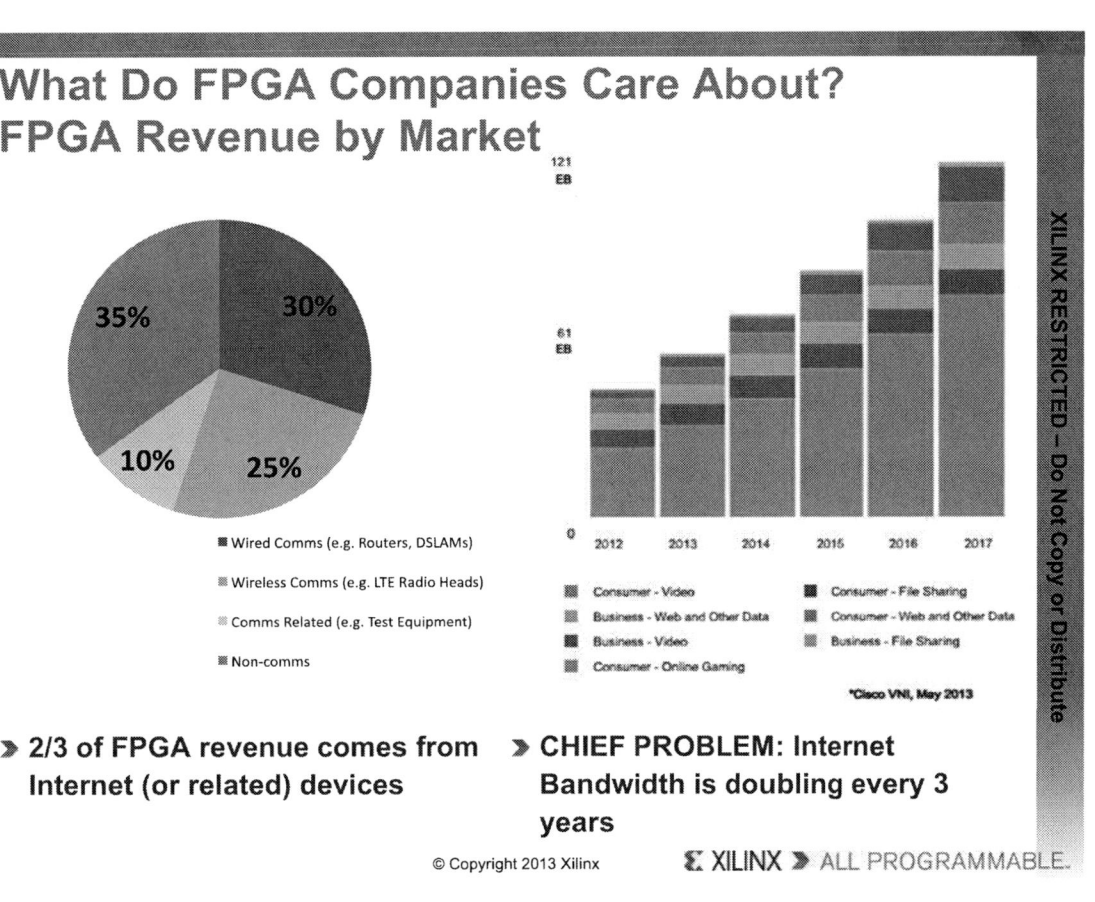

What Do FPGA Companies Care About? FPGA Revenue by Market

- Wired Comms (e.g. Routers, DSLAMs)
- Wireless Comms (e.g. LTE Radio Heads)
- Comms Related (e.g. Test Equipment)
- Non-comms

Consumer - Video
Business - Web and Other Data
Business - Video
Consumer - Online Gaming

Consumer - File Sharing
Consumer - Web and Other Data
Business - File Sharing

*Cisco VNI, May 2013

➢ 2/3 of FPGA revenue comes from Internet (or related) devices

➢ CHIEF PROBLEM: Internet Bandwidth is doubling every 3 years

© Copyright 2013 Xilinx

Wireless : Wireless Radio Heads

> **What does it do?:**
> - Trends:
>   - More advanced modulation techniques to approach Shannon Limit
>   - More, smaller radios (smaller cells)
>   - Severe cost/power limitations

> **Key Needs (2014-2015 customer prototypes) :**
> - Absolute lowest cost system solution
> - Move from 368MHz to 491MHz in low cost parts
> - Move from LVDS (IO) to JESD204b (GT) interface to DAC/ADC @ 12.5G
> - Shift to more GTs (maybe 20-50) and less IO (maybe 150) in small packages
> - Still needs lid (bare die not an option)

> **Key Needs (2016-2017 customer prototypes) :**
> - FinFET silicon running at core frequencies of 700MHz-1GHz
> - Need creative ways to wriggle out of extreme cost focus
>   - More Integration (e.g. ARM, MCM, fPLL, VCXO)
>   - Ideas for lower cost packages with many Transceivers and only a few IO (~100-150)

© Copyright 2013 Xilinx   **Σ XILINX ⟫** ALL PROGRAMMABLE.

Wired : Wired Routers and Traffic Managers

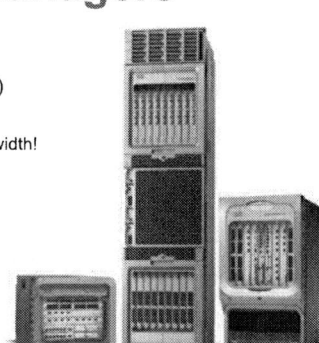

> **What does it do?:**
> - Aggregates, shapes, manages and prioritizes packet traffic (needs to buffer packets)
> - Trends
>   - More datapath bandwidth (Keep up with Optics) at lower power = More **Memory** Bandwidth!

> **Key Needs (2014-2015 customer prototypes) :**
> - More Memory Bandwidth at Low Power
>   - 400G of line bandwidth = 1.2T of memory BW (400G read, 400G write @ 66% efficient)
>   - Stacked memory is obvious solution, but is it ready? How much does it cost?
>   - Need efficient DDR4 solution too (covering our bases)
> - Serial Memories
>   - May have packages that are nearly all Transceivers! (lightweight)
> - Commodity 100GE (4x25.78125G) – 25-30G VSR in cheap packages
> - 25G backplane is a premium feature

> **Key Needs (2016-2017 customer prototypes) :**
> - 1T of line bandwidth needs 3T of memory bandwidth
>   - Stacked memory is the solution - Need lower cost + commercial
> - 25G backplane comes down in cost
> - Needs thermal relief help (70W+ in a 45mm package)
>   - Any ideas for better ThetaJA?

© Copyright 2013 Xilinx   **Σ XILINX ⟫** ALL PROGRAMMABLE.

Summary

» Package cost must stay low even with higher serial bandwidth
- Wireless:
  - Very, very low cost 12.5Gbps packaging for JESD204b (though fewer IO)
  - More on-package integration
- Wired:
  - Density: 100+ transceivers at 25G+ with very few IO (100-200)
  - Stacked Silicon
  - Thermal: How to get 50-100W with a 1" sink?

» Stacked Silicon is the way, but there are challenges
- Pros:
  - Order of magnitude less IO power per Tbps
  - Feasible: Xilinx in production on 2.5D stacked Si today (Virtex-7 FPGAs)
- Cons:
  - Business model built for consumer, not infrastructure

THE VISUAL COMPUTING COMPANY
MEPTEC ROADMAP 2013 – ABE YEE

GPU Roadmap

- GPU's continue on Moore's Law
- ~2X performance every 2 years
- Memory Bandwidth Critical
- GDDR5 - Kepler GPU - 1.5Ghz*4*384bit ->288Gbps
- Volta GPU generation -> 1Tbps
- Stacked Dram required to break memory wall
  - Requires 3D w/TSV
- Memory to GPU requires 2.5D w/TSV

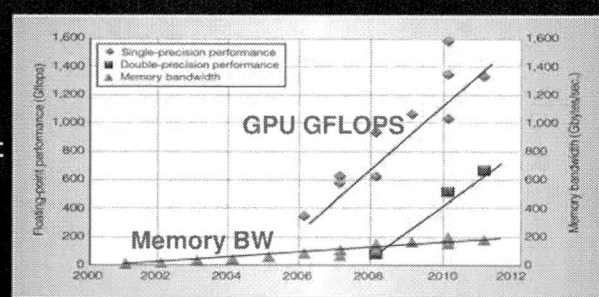

Figure 1. GPU processor and memory-system performance trends. Initially, memory bandwidth nearly doubled every two years, but this trend has slowed over the past few years. On-die GPU performance has continued to grow at about 45 percent per year for single precision, and at a higher rate for double precision.

Source: IEEE Micro Sept/Oct 2011

Mobile Roadmap

- **Current: LPDDR3 – 12.8 GBps**
  - 32bits*2 channels -> 64bits -> 8Bytes
  - 800Mhz*2 -> 1600Mhz
  - 8*1600 -> 12.8GBps
- **Future: LPDDR4 – 25.6GBps**
  - Clock 800->1600Mhz
- **Beyond: Will require WIO – 34GBps**
  - 64bits*4 -> 256bits -> 32Bytes
  - 32*533Mhz*2 -> 34GBps
- **Mobile will need 3D as well**

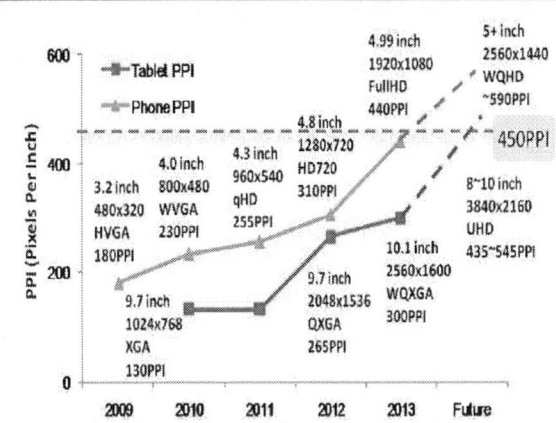

Mobile device display size and resolution trends

Source: Samsung, 2013 VLSI Symposium

Summary

- Visual Computing Requirements Continue on Moore's Law
- GPU has relied on GDDR5 but will require stacked memory for improved performance
- Mobile has relied on POP but will require stacked memory for improved performance
- 2.5D and 3D will be needed just around the corner

Panel 3 – Electrical Performance Requirements

Key Takeaways

- Electrical performance of package level interconnect is a big deal
- 2 Key Trends
  - Rise of serial bandwidth
  - Challenges of memory buffering
- How to get to the terabit internet
  - Backplane line rates double every two nodes
  - 1 Tb of line bandwidth needs about 3 Tb of memory bandwidth
- System architecture and partitioning critical
- Package design tools to enable non-EM engineer to visualize design performance
- Standards need to consider chip to chip
- GDDR5 and LPDDR4 run out of steam – 3D at 16nm for GPU, >28Gbps AP
- Shrinking memory supply base – impact on HBM/HMC supply

Ivor and John – Xilinx and Altera

- Transition of industry and importance of characterization engineering on packaging and device
- Interface between device and system
- Backplane line rate doubles every two nodes
  - How to get to terabit internet?
  - 10nm > 50Gb/s
  - 0Xnm > 100Gbps >
  - 50 – 100 entry point for mainstream optical switching
- Component power at 100W going to 200W in 2020
- Cost per bit converging regardless of bandwidth
- Circuitry density
  - Organic dry process and Si FBEOL converge in future
    - Not only Si and Glass are alternatives

Tom - Broadcom

- All markets are cost sensitive
- Technology innovation continues but....dark clouds on horizon such as the slowing down of Moore's law
- Questions
  - What portion of semiconductor industry success has been result of Moore's law
  - Are there alternatives to supplement the benefits of Moore's law
  - How much of the gap can be filled by these alternatives
  - What technical, organizational, etc resources are required?

- Tom's Thoughts
  - No single replacements
  - 3D TSV must be preceded by other 3D technologies
  - Winning strategies will rely on optimzation of partitioning, architecture and customization
  - SW needs to contribute not consume efficiency
  - Mixed signal and 2nd tier suppliers can be studied to understand how to survive without moore's law
  - Jonas law – swim fast or be eaten

Brad - Cadence

- Electrical assessment – early and often
- Standards are really a system challenge – chip to chip
- EDA can empower IC package designers with accurate electrical assessment eliminating iterations with analysis teams and little EM experience
- EPA modes
  - Signal analysis
  - Power/Ground Analysis
  - DC current analysis
- Graphical display at a push of a button – designer can make changes without iterative approach

Anthony - Xilinx

- 2/3 of FPGA revenue comes from Internet
- Internet bandwidth doubles every 3 years maybe as fast as every 2 years
- Trends
  - Rise of serial bandwidth to chips
  - Increase of buffering problems and need for memory bandwidth – DDR not keeping up
- Serial (Wireless)
  - More smaller cells
  - 500MHz in low cost parts
  - IO to
  - GT to DAC/ADC
  - FinFET to 700MHz - 1GHz
  - Lower cost many transceiver packages
- Buffering (Wired)
  - DDR not getting there
  - 400G of line bandwidth challenge for DDR
  - Stacked Si potential solution but how much and when?
  - Or more serial using fewer pins but latency and re-ordering challenges
  - Future 1T line bandwidth needs 3T of memory bandwidth
    - Stacked memory is solution

Abe - nVidia

- GPU's on Moore's law for 20 years
- Memory is like a big gasoline tank – needs efficient fuel injection system
  - Now trailing GPU performance
  - Utilizes a lot of space and impedes performance
- 16nm requires 1Tbps – cannot do with GDDR5
  - Require stacked memory – 3D
- In mobile, >25.6 Gbps LPDDR4 requires wide IO

Q&A

- Analog and mixed signal companies have competed
- Co-design between chip to package to system
  - Difficult also based on how companies are organized and communicate
  - Physical co-design progress but electrical co-design still challenging
- Megtron 6 and Megtron 4 adoption for wired applications as FR4 replacement
- Micron and Hynix as sole independent memory suppliers issue? Will they support HBM?

- Radiation tolerance at 16nm? Digitize and error correction
- Blurring the lines
- HMC vs HBM

Panel Four: The Importance of Industry Organization Collaboration

Moderator: Phil Marcoux, PPM Associates

Panelists:

- **CAMEST** – Dieter Bergman – IPC
- **EDPS** - Herb Reiter - eda2asic Consulting, Inc.
- **GSA** - Kenneth Potts - Cadence Design Systems
- **IPC** – Jasbir Bath - IPC
- **SEMI** – Paul Trio – SEMI

2013 MEPTEC
SEMICONDUCTOR ROADMAPS SYMPOSIUM

Session 4: The Importance of Industry Organization Collaboration

The cost of research and development for critical new technology in areas such as IC packaging and electronic assembly have stretched beyond the reach of many companies. This impacts the acceptance of promising new technologies and delays improvements in performance and cost. Recent technologies such as 3D are getting R&D help from the increasingly coordinated efforts of international consortia, and infrastructure to boost the successful technologies is coming from the major standards groups.

This session will consist of a panel of representatives from these organizations who will provide an update on their progress and discuss how they can work together to further assist the industry in developing and adopting advanced technologies.

Coalition for Advancement of MicroElectronic Systems Technology - "CAMEST"
Organization and Plans

Public Presentation
MEPTEC – September 24, 2013
Denny Fritz

Origin of CAMEST - 2000 JISSO
Mission Statement

The purpose of this council is to promote a strategic partnership among organizations interested in the total solution for interconnecting, assembling, packaging, mounting, and integrating system design and by increasing global awareness.

- *To accomplish these objectives, members will cooperatively work: to support standards development at a national or international level, to encourage the development of technology roadmaps, to address environmental issues, and to monitor market trends.*

- *These activities will be based on the principles of free enterprise, cooperation, and will be undertaken in a spirit of responsibility to the worldwide electronics industry.*

Forming Coalition for Advancement of MicroElectronic Systems Technology "CAMEST"

- Mission Statement: *"CAMEST" is dedicated to the identification and dissemination of the critical technology application knowledge needed for the further development of the electronics industry. This organization identifies gaps in design, manufacturing, test and reliability across all aspects of electronic component assembly and subsystem manufacturing from semiconductor to final assembly, and facilitates cooperation among industry, academia, government and existing consortia to deliver solutions*

"CAMEST" Goals

1. Serve as center for the gathering and distribution of technical information relevant to stakeholders

2. Identify gaps not publically recognized

3. Expedite transfer of knowledge of needs and coordinate effort to bridge barriers which may arise between stakeholders.

4. Foster cooperation between industry, academia, government stakeholders and existing consortia to accelerate the delivery of viable solutions

"CAMEST" Coalition Status

- Incorporated in Delaware
- Holds regular weekly meetings
- Set an advisory board
- Have worked on by-laws
- Started first "technology map" – Multichip Packaging. Will be evergreen – updating status and actions needed, more often than roadmaps.
- Seeking to obtain cooperation of individuals, OEMs, consortia, universities, government

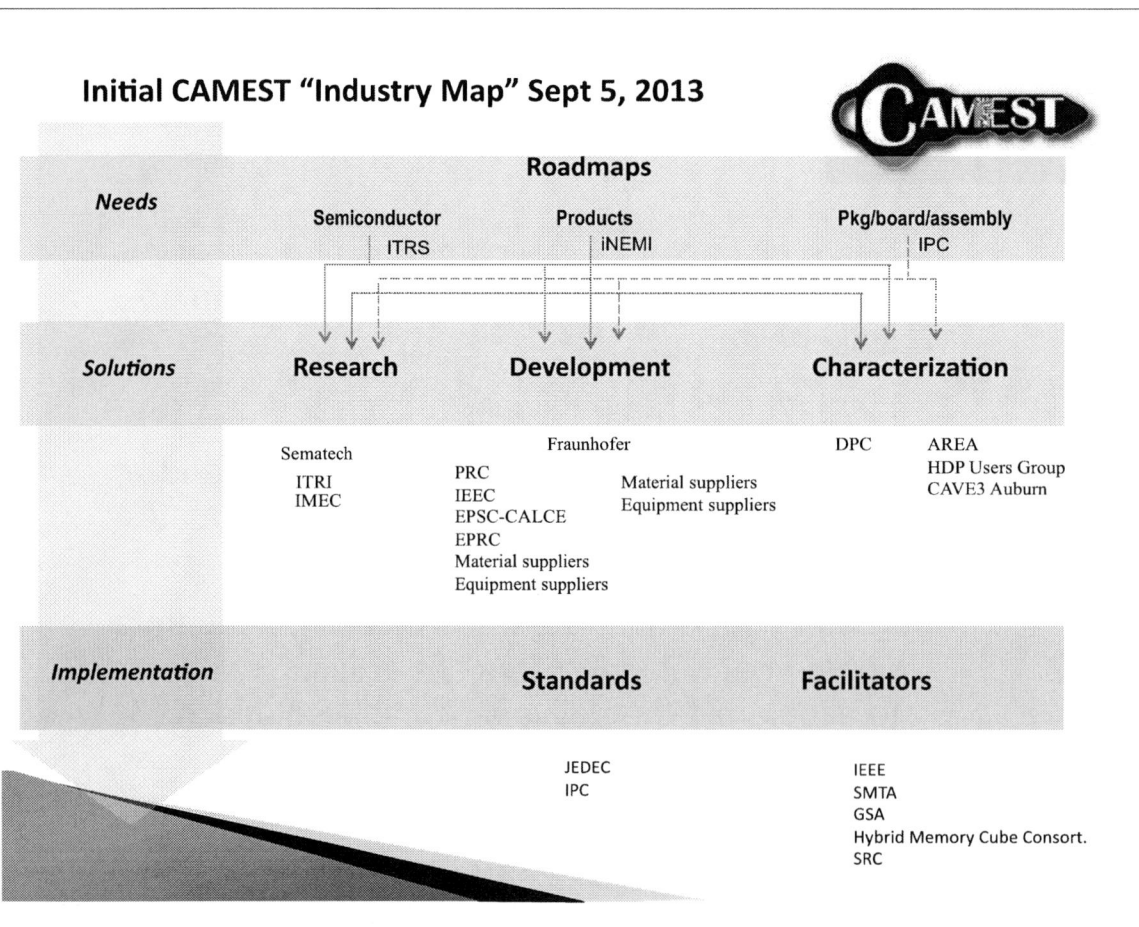

CAMEST Potential Affiliations

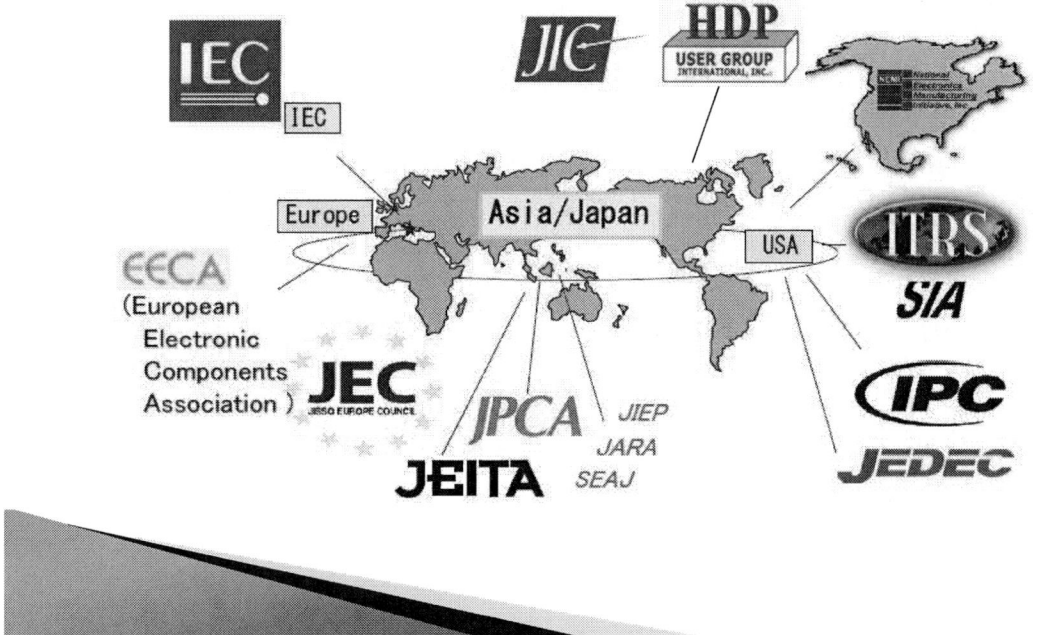

CAMEST Gap Analysis
Projects by other associations Sept, 2013

- High Density Package Users Group – HD PUG Current Projects – 2014 planning meeting coming.

Semiconductor Packaging			
Polymer Ball Interconnect	Brian Smith	Oracle	Implementation
Process Sensitive Components	Laurence Schultz	IBM	Implementation
Counterfeit Electronics	Laurence Schultz	IBM	Implementation
FCBGA Package Warpage	Robert Smith	Intel	Definition
Emerging Technology			
Through Silicon Vias (TSV)	Larry Marcanti	Oracle	Idea

- AREA (Universal Consortium). Next members meeting starts September 25.
    - APD2A: Fine Pitch Project Proposal
    Author: Michael Meilunas
    The primary objectives of this project are to
    1.) Determine optimum design and processing parameters for successful 0.3mm pitch assembly and reliability.
    2.) Evaluate the effect of plated vs. non-plated via-in-pad features using reliability tests.
    3.) Study thin die components.

CAMEST Gap Analysis
Projects by other associations Sept, 2013

- **Center for Advanced Vehicle and Extreme Environment Electronics – Auburn University**
  - **Flip Chip and Underfills**
- In this research area, materials and processes are being explored for flip chip on laminate, flip chip BGA packaging, CSP (redistributed die, Ultra-CSP, etc.) assemblies deployed in extreme thermal cycling environments. The primary objective is to develop a fundamental understanding of the reliability of flip chip applications in harsh environment applications and High End Microprocessor Packaging. Study next-generation materials (Nano-structured underfills, High-Reliability STABLCOR Substrates, Thermal Interface Materials, Chip-Level Interconnects). Project deliverables include design and material guidelines for flip chip packages used in the automotive thermal cycling environment; material properties and adhesion characteristics of underfill encapsulants; flip chip thermal cycling reliability data; assembly and manufacturing processing recommendations; and finite element and material models for application to future package designs.

CAMEST Gap Analysis
Projects by other associations Sept, 2013

- **Georgia Tech Packaging Research Center (PRC). Georgia Tech *Global Interposer Technology (GIT)* conference – November 17-19.**
  - 5. INTERCONNECTIONS, ASSEMBLY, & RELIABILITY
    Research Focus: Cu-to-Cu low temperature bonding with adhesives, low temperature and ultra-fine pitch metallurgical bonding for Cu interconnections, underfill processes for low stand-off, chip-last embedded IC interconnections, highly reliable second level interconnections for low CTE packages, nano-structured interconnections.
  - Research Targets:
    - Chip-to-Package I/O pitch – 10-50µm array
    - Bonding Temperature – <200°C
    - Package-to-PWB I/O Pitch – 300-500µm

eda2asic

MEPTEC Roadmap Symposium
The Importance of Industry Organization Collaboration

Herb Reiter, eda2asic Consulting, Inc.
herb@eda2asic.com, 1-408-981-5831
Biltmore, Santa Clara, CA
September 24, 2013

9/29/13

1

---

eda2asic Electronic Design Process Symposium

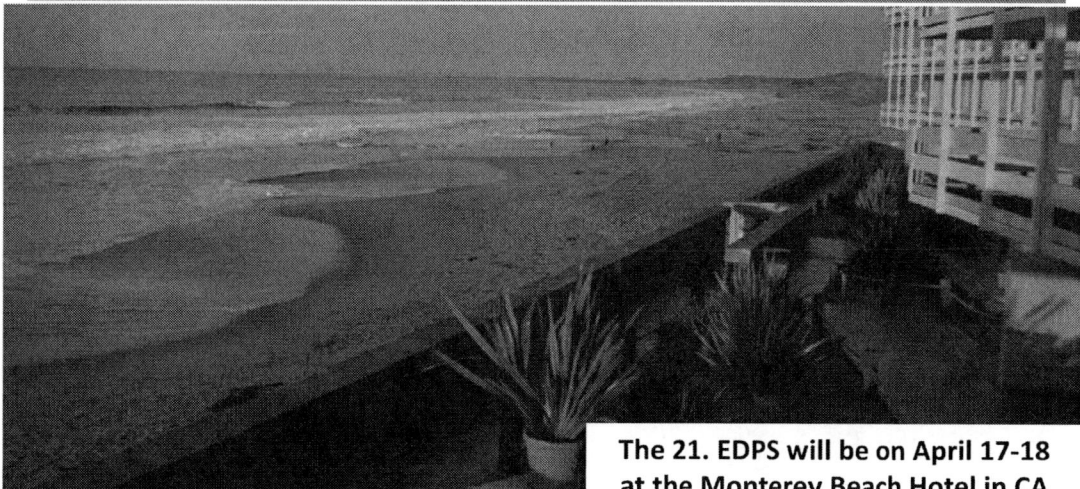

The 21. EDPS will be on April 17-18
at the Monterey Beach Hotel in CA
http://www.eda.org/edps/

Thursday, 4/17, 9 am to 6 pm: Several very relevant topics presented & discussed
Friday, 4/18, 9 am to 3 pm: Entire team concentrates on ONE highly relevant topic

9/29/13

2

eda2asic — EDPS Value Proposition

A group of System-, EDA- and Semiconductor experts meets every April in Monterey, CA, to address important current and future industry needs and to outline ideas for practical solutions.

+ Relatively small team (~50 people) allows in-depth discussions and high-quality networking with other technology drivers
+ Experienced, engaged and open-minded organizers
+ Thursday: Keynote and 3 to 4 focused sessions; dinner
+ Friday: ONE topic for keynote, presentations and panel
+ New presenters welcomed –> please show your interest soon
+ Passionate panel discussions with lots of audience participation
+ Only ~ 60 miles away from Silicon Valley's hectic and stress
+ Early Bird registration fee and hotel rooms reasonably priced
+ The – so far – best kept secret in our industry

9/29/13

---

eda2asic — Examples for Topics @ EDPS

2013

- Thursday, April 18:
  - Xilinx Keynote
  - ESL & Platforms
  - Design Collaboration
  - 3D-IC Update
  - Networking Dinner
- Friday: **FinFETs**
  - Keynote by Semiwiki
  - IP & Tools Challenges
  - Panel: Foundry Challenges

2012

- Thursday, April 5:
  - Altera Keynote
  - System Design
  - Cloud Computing
  - Low Power Design
  - Networking Dinner
- Friday: **2.5/3D-ICs**
  - Keynote by Qualcomm
  - Design Tools
  - Panel: 3D Ecosystem

9/29/13

eda2asic   **Companies Presenting @ EDPS**

2013 - FinFET
- Adapt IP
- Alcatel-Lucent
- ARM
- Cadence
- Docea Power
- eda2asic
- E-System Design
- Gary Smith EDA
- Globalfoundries
- Intel
- Mentor
- Micron
- NetApp
- Nimbic
- Oracle Labs
- Qualcomm
- Semiwiki
- Space Codesign
- Swan on Chips
- Synopsys
- TSMC
- Xilinx
- Xuropa

2012 – 2.5/3D-IC
- Altera
- AMD
- ARM
- Broadcom
- Cadence
- eda2asic
- Gary Smith EDA
- Intel
- Mentor
- Nimbic
- PPM Associates
- Qualcomm
- Rambus
- Swan on Chips
- Synopsys
- Tensilica
- TSMC
- Univ. Wisconsin
- Vista Ventures
- Xuropa

See you at EDPS 2014 -- on April 17 & 18

9/29/13

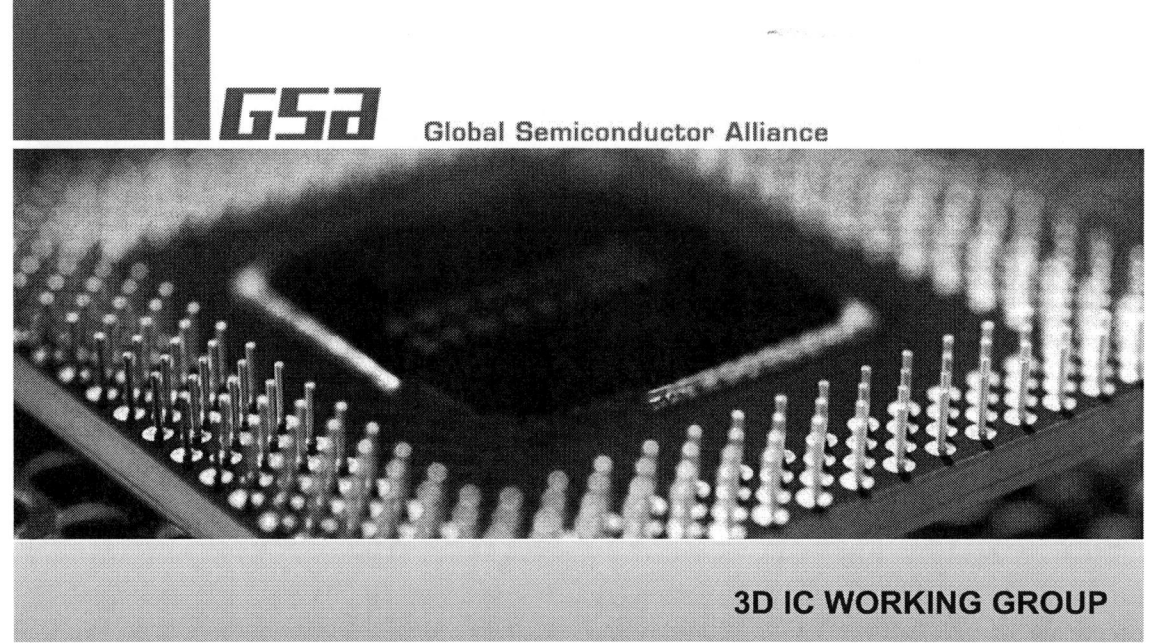

Global Semiconductor Alliance

3D IC WORKING GROUP

KEN POTTS
3D IC WORKING GROUP CHAIR

cādence

3D IC Working Group Objective

- Promote adoption and proliferation of 3D-IC technology
  - Gather business intelligence in the areas of supply chain and technology
  - Provide access to information to support members in their business decisions
    - Technology maturity and adoption
    - Implementation best practice
    - Business model options

cādence

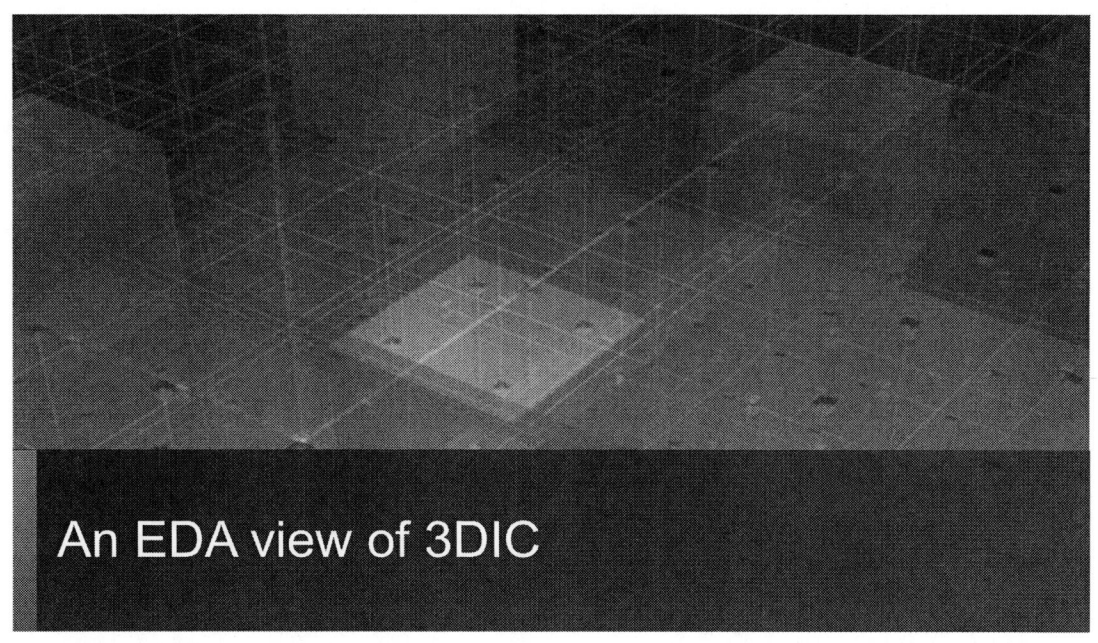

An EDA view of 3DIC

Ken Potts
Group Director, Strategic Marketing
Cadence Design Systems

cādence

3D-IC is a bridge for "More Than Moore" solutions

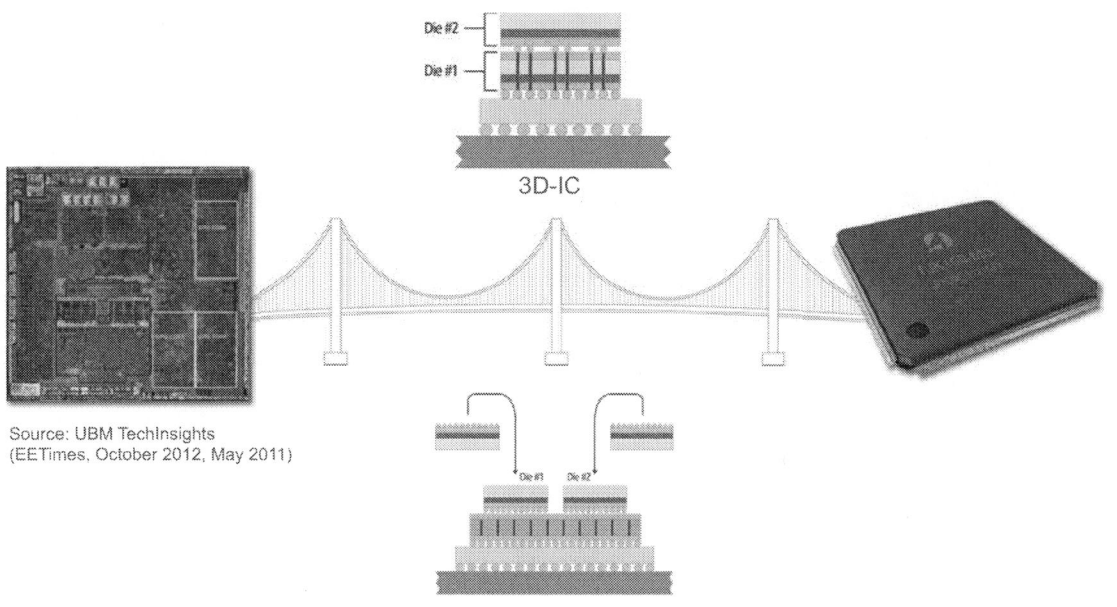

Source: UBM TechInsights
(EETimes, October 2012, May 2011)

cādence

Why 3D?

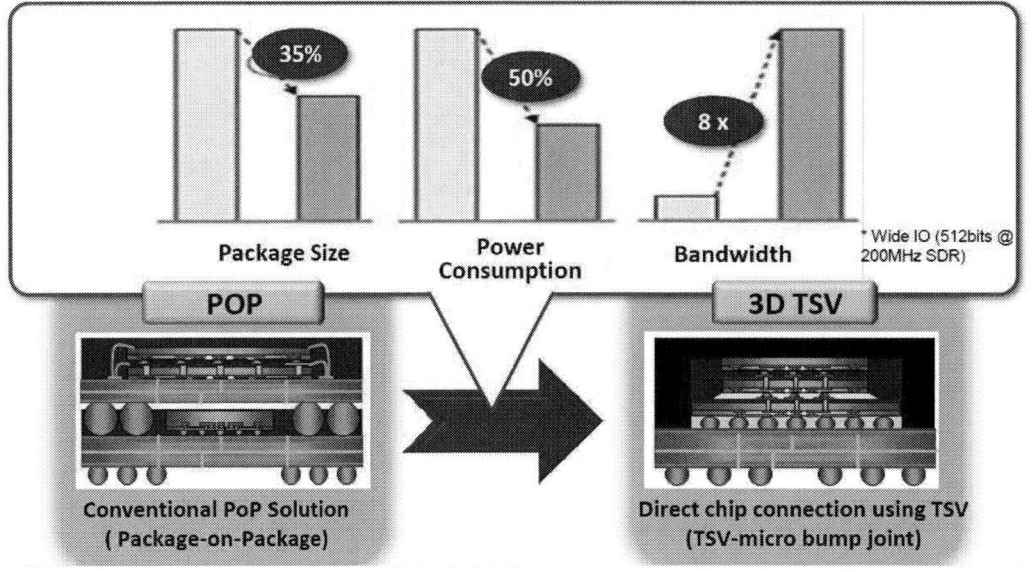

Source: Samsung Electronics 2012

Short-, medium-, and long-term path to 3D-IC
EDA work starts at least 3-4 years earlier

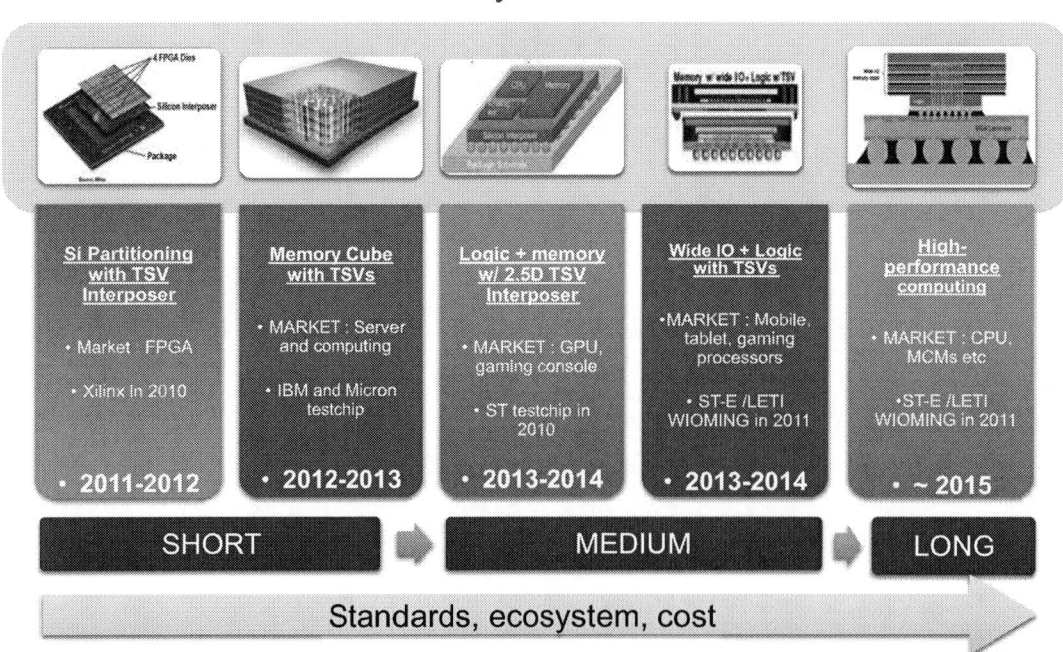

3DIC Design Flow Challenges

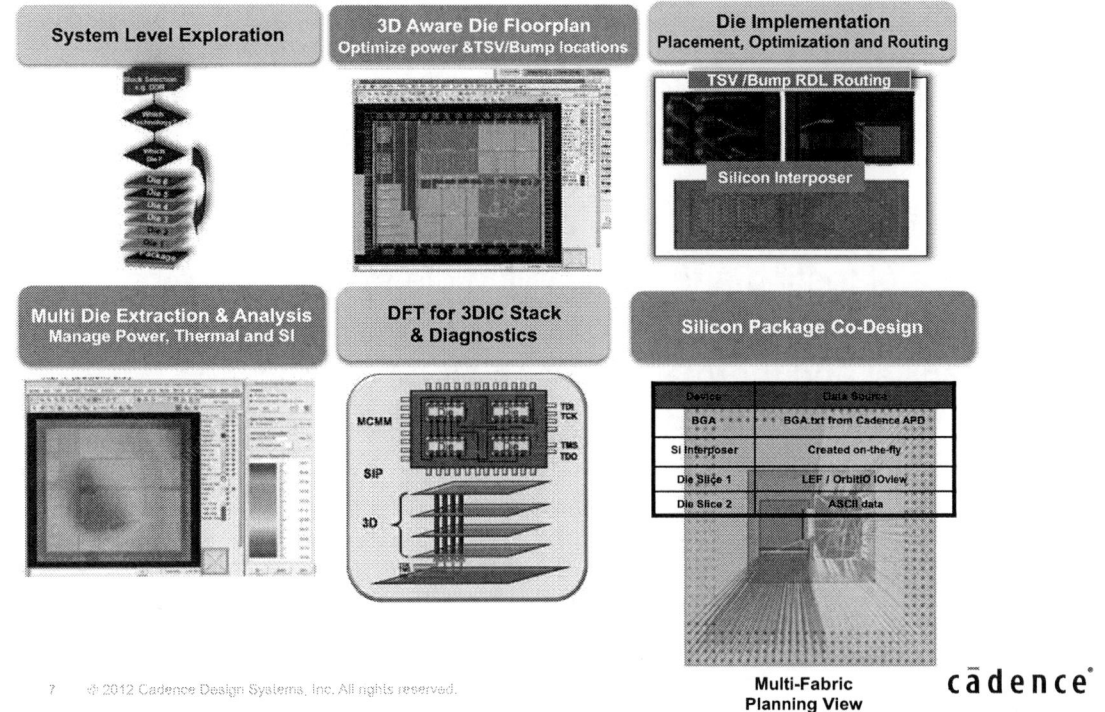

Multi-Fabric
Planning View

cādence

EDA for 3D-IC: current status

- Some designs can be done with existing tools
- More sophisticated implementation and analysis tools are being developed
  - Examples shown before
- Gaps exist
  - Path-finding
  - System analysis and optimization with implementation details
- No fundamental limitations: economics dictate rate-of-tool development

cādence

3D integration challenges

- Economical 3D stacking in high-volume manufacturing presents many challenges
- Benefits must exceed the additional costs of TSVs, and yield fallout
- Logistics of testing and assembling die from multiple sources can be substantial
- Mechanical and thermal issues must be addressed

cādence

Collaboration is key to success
3D-IC ecosystem and collaborations

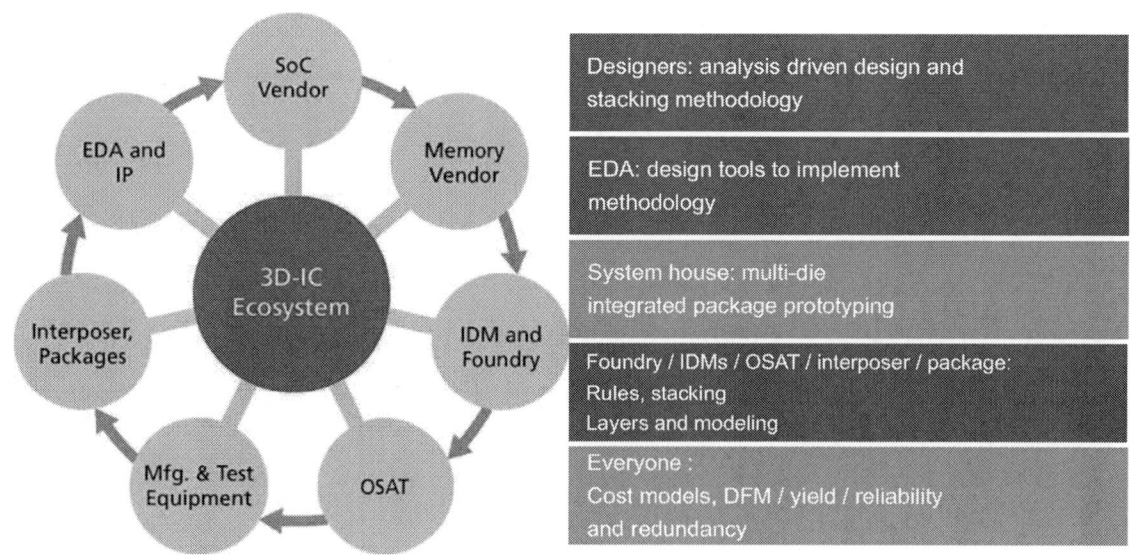

cādence

cādence®

The Importance of Industry Organization Collaboration

Jasbir Bath
Principal Engineer, Assembly Technology
IPC
Email: JasbirBath@ipc.org

Association Connecting Electronics Industries

IPC

- Founded in 1957 to meet Industry Needs with focus on Design, PCB Manufacturing and Electronics Assembly.

- Organization develops standards in Electronic Components, Printed Circuit Boards, Electronics Assembly and Design.

- IPC currently has over 260 standard subcommittees and task groups.

- IPC has a network of over 10,000 people who volunteer for standards committee work.

- Over 3,000 member companies globally.

Association Connecting Electronics Industries

The "New" Industry Reality: Overview

- Few vertically integrated companies remaining.
- "Company-based" research, development & technology pool reduced
- Needed knowledge is distributed across many countries, many companies.
- "Lean" operating staff and short-term ("quarterly results") focus reduces availability of some resources
- Resulted in sharing of Experts/Inter-organization co-operation for areas such as Technology Roadmap Development (IPC, INEMI, ITRS etc)
- <u>Industry organizations will increasing have to work together/collaborate to achieve benefits for members.</u>

Association Connecting Electronics Industries

Example: 2.5D/3D Packaging Priorities and Gaps

- To address current technical issues and infrastructure challenges for the development of 3D-IC packaging, a coordinated and collaborative effort is needed in the supply chain.
- Organizations including IPC, SEMI, JEDEC and GSA are looking to organize industry discussions and help in the development of standards in this area as needed.
- Surveys currently being conducted to understand where standardization would benefit 3D packaging development.

Association Connecting Electronics Industries

SEMI Standards Program Overview

The Importance of Industry Organization Collaboration
MEPTEC Semiconductor Roadmaps Symposium 2013

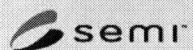

MEPTEC Semiconductor Roadmaps Symposium 2013

About SEMI and SEMI Standards

SEMI

- Global industry association serving the manufacturing supply chain for the micro- and nano-electronics industries, including:
  - Semiconductors | PV | HB-LED | FPD | Printed & Flexible Electronics | Other Emerging Markets
- Member Services
  - Expositions | Industry Research & Statistics | Technical Programs | Standards
  - Executive Conferences | EHS | Advocacy | Special Interest Groups

SEMI Standards

- Established in 1973
- Experts from the microelectronic, display, photovoltaic, and related industries
- Develop globally-accepted, consensus-based technical standards
- We are international
  - United States | Japan | Europe | Taiwan | Korea | China

MEPTEC Semiconductor Roadmaps Symposium 2013

SEMI Standards Activities

Overall:
- 23 Technical Committees
- 200+ Task Forces
- 400+ Activities
- 880+ Published Standards
- 4500+ Program Members

- 3DS-IC
  - 5 Published Standards | 7 Task Forces
  - Current activities
    - Material Properties for 300 mm Wafer Stack | Bond Void Measurements
    - Alignment Marks | CMP & Microbump Processes for Frontside TSV Integration
    - Thin Chip (Die) Bending Strength Measurement

- 450 mm Wafers
  - 18 Published Standards | 13 Task Forces
  - Current efforts
    - 450 mm Notchless Wafers | Wafer Shipping System

- EHS
  - 29 Published Standards | 29 Task Forces
  - Current activities
    - Seismic Protection | Hazardous Energy Isolation | Ergonomics | Energy Conservation
    - Chemical Exposure | Non-ionizing Radiation | Fire Protection

MEPTEC Semiconductor Roadmaps Symposium 2013

Critical Role of Standardization

Reduce Manufacturing Complexity, Allowing Companies to Focus on Innovation

Customer ⟷ **Supplier**

Customer		Supplier
Multiple suppliers reduce costs	⟷	Productive spare parts inventories
Multiple suppliers improve quality	⟷	Focus on critical performance variables
Faster supplier response	⟷	Faster deliveries
Shorter time to market	⟷	More resources for R&D

Call to Action:
- Understand the value of Standards
- Participate in standards development efforts
- Raise awareness of ongoing standardization activities and encourage others to join

MEPTEC Semiconductor Roadmaps Symposium 2013

MEPTEC Technology Roadmap 2013
- Session 4 transcription

Santa Clara, CA

Sep 24 2013

Session 4: Importance of Industry Organization Collaboration

Session Leader: Phil Marcoux, PPM Associate

Transcription: Hongxia Sun, Bernard Adams, STATSChipPAC Inc.

Panelists:

- Dieter Bergman, Director of Technology Transfer, IPC
- Herb Reiter, eda2asic
- Kenneth Potts, Group of Director of Strategic Planning, Cadence Design System
- Jasbir Bath, Principal Engineer, IPC
- Paul Trio, Senior Manager of North America Standards Operations, SEMI

Outline

- Session 4 introduction
- Organization updates
- Key takeaways
- Discussions on challenges and solutions
- Other notes

Session 4 Introduction

As emerging advanced technology continues to push for performance improvement with cost reduction, there are lots of activities for research and development in areas such as IC packaging and electronic assembly that have stretched beyond the reach of many companies. This impacts the acceptance of promising new technologies and delays improvement in performance and cost. Therefore, collaboration across organizations is critical and required.

Given the above, panelists in session 4, representing 5 different organizations, provided updates and lead discussions on their progress to assist the industry in developing and adopting advanced technologies.

Organization updates:
CAMEST - Coalition for Advancement of MicroElectronic System Technology

- Panelist 1 representing CAMEST –Dieter Bergman.
- CAMEST was founded in 2013 to further advance packaging solutions. 7 volunteer members with 65 people expressing interest. Purpose of this council is to promote strategic partnership among organizations interested in total solutions for interconnecting , assembling, packaging, mounting and integrated system design by increasing global awareness.
- Introduced CAMEST which has incorporated organizations and individuals.
- Goal to get roadmaps, come up with solutions and help implementation.

Organization updates:
EDPS - Electronic Design Process Symposium

- Panelist Herb Reiter from eda2asic – representing EDPS
- Announced coming EDPS event in Apr 2014. Topics discussed in 2013 were low power, 2.5/3D packaging technology, etc.
- Introduced EDPS's value proposition, topics at EDPS and companies presenting at EDPA.

Organization updates:
GSA – Global Semiconductor Alliance
3D IC WORKING GROUP

- Panelist Ken Potts from Cadence – representing GSA

- Introduced 3D IC Working Group's organization objective of promoting adoption and proliferation of 3D IC technology.

- Provided an EDA view of 3D IC with discussions such as "Why 3D?", "Path to 3D" and challenges including current status and collaboration for the "key to success".

Organization updates:
IPC – Association Connecting Electronics Industries

- Panelist Jasbir Bath – representing IPC

- IPC , an organization with sources for industry standards, training, market research and public policy advocacy, with 260 standard subcommittees and task groups, Over 3,000 member companies globally.

Organization updates:
SEMI Standards Program Overview

- Panelist Paul Trio – representing SEMI

- Introduced SEMI's services, standards and activities, as a global industry association serving the manufacturing supply chain for micro- and nano-electronics industries.

- Discussed critical role of standardization to reduce manufacturing complexity to allow companies to focus on innovations etc.

Key TakeAways

1. Five Organizations driving Collaboration and Standards. CAMEST, EDPS, GSA, IPC, and SEMI.
2. Benefits of Successful Collaboration – Drives Supply Chain, enables cost reductions, and improves quality.
3. Dieter Bergman presented a clear definition of a "Good Standard".
   1. Can be included in a contract
   2. Is clear
   3. Sets out requirements that can be measured.
   4. Addresses Quality and Reliability
4. Timing – Collaboration Occurs when multiple Industry Players are willing to share Best Known Methods and when a common solution is no longer driving competitive gross margins or a Disruptive Event occurs (Phil Marcoux related a call from the Department of Defense initiated the industry to address the SMT issue).
5. Collaboration to date on core issues such as the 3DIC business model is stalled. Ken Potts, GSA, related that there is interest with the industry players in vetting the business model but at this point, each individual player is focused on local optimization. Jaspir Bath reported a survey released to industry had little feedback so far.
6. 3D – IC Technology is now. Abe Yee from NVIDIA related that their graphic processor unit is now stacking memory. China has invested $100M to develop the technology in Shanghai.
7. There are no fees to becoming a SEMI Standards Member. Standards are available for purchase from SEMI which helps fund the Standards Program.

Discussions on Challenges and Solutions

Two hot spots in panel discussion:

1. Organizational collaboration
2. Emerging technology & standardization

Discussion: Collaboration Importance

1. Promote new technology for acceptance
2. Bandwidth to drive activities to benefit industry with people committed to work and spend their time on this effort.
3. Help to know and create business model with collaborations, verify supply chain and understand direction of product.
4. Intelligence on market and technology. Organizations represent different view of product, get different perceptions to think from the overview point to understand characteristics and know where to go and what to do.
5. Collaboration is key to success – example of 3D IC ecosystem & collaboration.
6. Diversity, e.g., through collaboration customers working on 2.5D also look at what's going on 3D technology, and vice versa.

Discussion: Emerging Technology & Standardization

1. Critical role of standardization and call to action to understand and participate in standard development efforts and raise awareness of ongoing activities etc.
2. How to manage consistency of standards. Standards from different organizations; organizations need coordination and come up with joint standards and then each organization promotes to their members.
3. New reality for new knowledge, distribute new technologies & concepts – example of 2.5D/3D packaging priority and their gaps.
4. Promote adoption of 3D IC technology with organizational collaboration , including topics such as why 3d, short term and long term aspects, and 3D standardization etc.
5. Joint work to reduce standards overlap and over saturation.
6. Time-line concerns for updating progress and update of standards, including discussions regarding 450mm & 300mm technologies. Consensus building can take time for high-value standardization activities and when several key stakeholders are involved. End user participation and strong leadership are key.

Other Notes

Session 4: The Importance of Industry Organization Collaboration

Session Leader: Phil Marcoux, PPM Associates
Panelists:

- CAMEST –Dieter Bergman. Founded in 2013 to further advance packaging solutions. 7 volunteer members with 65 people expressing interest.
- EDPS - Herb Reiter, eda2asic . Facilitates industry collaboration through a yearly meeting. Next meeting is April 17-18, 2014 in Monterrey, CA. Past years topics were 3D-IC and Finfet Technologies. 50-60 attendees normally attend.
- GSA - Kenneth Potts, Discussed the role of GSA and the working group for 3D-IC 's , which is founded for ecosystem development, information dissemination, and dedicated to the acceleration and commercialization of 3DIC Stacking Technology. There are many other working groups such as analog mixed signals, IP, MEMs, and Supply chain performance.
- IPC – Jasbir Bath, Principal Engineer, founded in 1957 to promote design, PCB manufacturing and electronic assembly. 260 subcommittees.
- SEMI – Paul Trio, Senior Manager of North America Standards Operations, SEMI. International organization founded in 1973 for the semiconductor manufacturing supply chain. Five published 3D-IC standards published with seven active task forces.

AUTHOR INDEX

Ahmad, Mudasir...1, 13
Barber, Ivor 56, 70
Bath, Jasbir...75, 90, 94
Bergman, Dieter 75, 76, 94
Crisp, Richard 19, 21, 39
Delacruz, Javier................................19, 27, 39
Demmin, Jeffrey C. 19
Fritz, Denny 76
Ghahghahi, Farshad................................. 4, 13
Gregorich, Tom 56, 58, 70
Griffin, Brad 56, 62, 70
Huemoeller, Ron 19, 30, 39
Kinsman, Larry 7, 13
Marcoux, Phil 75, 94
Master, Raj 19, 33, 39
Matthew, Linda 39
Potts, Kenneth..................................75, 84, 94
Reiter, Herb 75, 81, 94
Savala, Karen.. 44
Shangguan, Dongkai 19, 36, 39
Strothmann, Tom 9, 13
Torza, Anthony....................................56, 65, 70
Trio, Paul 75, 92, 94
Upadhyayula, Suresh 11, 13
Werbaneth, Paul.. 13
Yee, Abe 56, 68, 70

MEPTEC
315 Savannah River Road
Summerville, SC 29485

ISBN 978-1-62993-805-9